# The Political Economy of Science

## Ideology of / in the Natural Sciences

Edited by

Hilary Rose

and

Steven Rose

*First published 1976 by*
**THE MACMILLAN PRESS LTD**
*London and Basingstoke*
*Associated companies in New York Dublin*
*Melbourne Johannesburg and Madras*

ISBN 0 333 21138 3 (hard cover)
0 333 21139 1 (paper cover)
Typeset in Great Britain by Reproduction Drawings Ltd
**Printed in England.**

To the heroic peoples of Indochina, who
demonstrated to the world how to struggle successfully
against the science and technology of profit and
oppression

# Contents

# Acknowledgements

The history of these two volumes: *The Political Economy of Science* and *The Radicalisation of Science* with their common sub-title, *Ideology of/in the Natural Sciences*, is told in the Introduction which is reproduced in both books. Here, as Editors, we should like only to express our thanks to those who have helped in the creation of this collective endeavour. We must begin by thanking the contributors as individuals, who have not only given of their time and work freely in their own writing but also in their criticism and discussion of drafts of one anothers' chapters. Many of the chapters have benefited from being discussed at meetings within the polytechnics and universities and of trade unions and community groups. For those of us who work in the education or research systems the rediscovery of the sociology of scientific knowledge has provided a stimulating and critical environment for discussion of the concerns of these books. But it is to the radical science movement that our debts are greatest. To cite all the groups and individuals who have criticised and unstintingly aided the development of this work would be to give almost a history of the radical science movement. However, we should mention in particular: in Britain, *Science for People*, the Women and Science collective, *Radical Science Journal*, the Campaign on Racism, IQ and the Class Society, *Radical Philosophy* and the Indochina Solidarity Conference; in France, *Impascience*; in Italy the Science for Vietnam collectives; and in the United States the Science for the People groups in Minneapolis, Chicago, Boston and New York.

Many people have helped by reading and commenting on particular chapters, and we would here particularly mention: Francis Aprahamian, Pat Bateson, Paul Emmerson, Dot Griffiths, John Hambley, Altheia Jones, John Marriott, Ralph Miliband, Charles Posner, Jerry Ravetz, Ken Richardson, Fred Steward, Richard Whitley and Ron Wilson. Others have provided political and intellectual support over the period of production of the books — sometimes more than they may themselves have realised — Mary Evans, Mike Faulkner, Nora Frontali, Luke Hodgkin, Ian Muldoon, Helga Novotny, Felix Pirani, Esther Saraga, Tim Shallice, Joe Schwartz, Paul Walton, John Westergaard, Maurice Wilkins,

Charlotte Wolfers and John Wolfers. The help received from other friends and colleagues is acknowledged in various footnotes in the text. Janet Hackett, Liz Hainstock and Mac Foxley typed and copied many of the drafts; Laurie Melton chased endless references. Paulette Hutchinson translated the chapter by Jean-Marc Lévy-Leblond, 'Ideology of/in Contemporary Physics', and E. Maxwell Arnott translated Liliane Stéhelin's 'Sciences, Women and Ideology'.

We would also like to thank those editors and publishers who gave permission for already published material to be used. The publishing history is as follows: An earlier version of 'The Incorporation of Science' appeared in *Annals of the New York Academy of Science*, 260 (1975) pp. 7-21. 'The Production of Science in Advanced Capitalist Society' is to be expanded to book-form and published, in Italy, in 1976 by Feltrinelli. 'On the Class Character of Science and Scientists' first appeared in *Temps Modernes*, 29 (1974) pp. 1159-77. 'Contradictions of Science and Technology in the Productive Process' is based on earlier lectures at AUEW—TASS summer school in July 1972, the British Computer Society's Social Responsibility Group in November 1972, the Oxford Society for Social Responsibility in Science, March 1973 and the Conference *Is there a Socialist Science?* February, 1975. Earlier versions of 'The Politics of Neurobiology' were published as 'Do not adjust your mind, there is a fault in reality', in *Social Processes of Scientific Development*, ed R. Whitley (London: Routledge & Kegan Paul, 1974) and *Cognition*, 3 (1974) pp. 479-502, 'Scientific Racism and Ideology' draws on S. Rose, J. Hambley and J. Haywood, 'Science, Racism and Ideology', in *The Socialist Register* (London: Merlin Press, 1973) and also S. Rose, 'Scientific Racism and Ideology', in *Racial Variation in Man*, ed. F. J. Ebling (London: Blackwells for the Institute of Biology, 1975). An earlier version of 'Womens Liberation: Reproduction and the Technological Fix' appeared in *Sexual Divisions and Society*, ed. D. Barker and S. Allen (London: Tavistock Press, 1976). 'A Critique of Political Ecology' originally appeared in *Kursbuch*, 33. The translation is by Stuart Hood and appeared in *New Left Review*, 84, (1974) pp. 3-31. We are grateful to the translator and the editors of *New Left Review* for permission to reproduce it here. 'Science, Technology and Black Liberation' is revised from an original article which appeared in *The Black Scholar*, 5 (1974) pp. 2-8. 'History and Human Values: a Chinese Perspective for World Science and Technology' was originally a lecture for the Canadian Association of Asian Studies, Montreal, May, 1975, published in *Centennial Review*, 20 (1) (1976) pp. 1 - 35. We thank the Editor for permission to reproduce it, with minor editorial changes, here. 'The Radicalisation of Science' first appeared in *The Socialist Register* (London: Merlin Press, 1972) and was subsequently reprinted in *Science for People*, 21 and 22 (1974).

# Introduction

In 1971, we began to discuss the idea of collecting together material for books on the theme of ideology of/in the natural sciences with other activists in the radical science movement. The response was positive and unequivocal. We could see that the political struggles in which the movement was engaged, beginning in different ways in the various advanced capitalist countries — the Indochina war and pollution in the United States, Britain, Japan and Australia; the hierarchy and elite nature of scientific practice in France and Italy — nevertheless were moving towards a series of fundamental questions which underlay them all. Scientists who had begun by feeling that 'their' science had been betrayed in the defoliation campaign in Vietnam, or that 'their' scientific community was a hollow myth, began to ask such questions as: Whose science is it? Who pays for it? Who decides it? Who benefits from it?

Because the production system of science requires interaction between workers at the international level, through journals, conferences, research centres, and so forth, concerns and issues which were felt in one section of the system rapidly spread and were taken up elsewhere. (In practice the movement had to learn that the vaunted internationalism of science was a function of its mode of production, just as much as contemporary capitalism demands the existence of the multinational corporation.) None the less, the differing political traditions — Marxist in France and Italy, social democrat in Britain and populist in the United States — meant that problems were seen and articulated in different ways. In France and Italy following 1968 there were laboratory occupations and attempts to develop self-managing scientific collectives involving the workers in particular institutes. In Britain, the campaign against chemical and biological warfare developed in pressure-group style, with attempts to use the media, ask parliamentary questions, persuade trade-union branches to pass resolutions, and urge moral renunciation of CBW work on individual scientists. In the United States, the work of the Honeywell collective centred on raising the consciousness of workers at Honeywell plants, designing and making fragmentation weapons for use in Vietnam, concerning the

nature of their product. In Japan, the campaign around mercury poisoning at Minamata involved grass-root mobilisation amongst communities
directly at risk from the pollutant. Some of these struggles were more
politically advanced than others, and even in any given situation there
were confused and contradictory ideas as to both the over-all strategy
of the campaign and the immediate tactics involved. Political action has
taken place in many different areas: within the scientific occupation
itself; in conjunction with factory workers; with local communities;
and in support of liberation struggles.

However, particularly in the United States and Britain, countries
with the most developed scientific production systems, and hence the
most organised scientific movements, these movements have been slow
to develop a theoretical perspective which would enable them to articulate the links between struggles in the different areas. While particular
groups have focused on, for instance, issues of the computer invasion of
privacy, or alternative technologies, there was little clarity about the
goals of the movement: was it to secure international law on CBW, to
unionise or to radicalise scientists, to aid workers in their struggle against
pollution in the work-place, or to act as a focal point in the general
struggle to overthrow the capitalist system? Instead, a cheerful and
energetic eclecticism prevailed. Initially, this was a strength, as new
spaces for action were found — spaces, it is important to say, that were
deemed not to exist by the old left orthodoxy — but by the early 1970s
most activists were recognising the practical and urgent need for theory.
They recognised that it was time to move beyond the early pragmatic
phase to a stage at which the contradictions present within science
could be seen as part of a general revolutionary perspective. This meant
not only strengthening the movement's understanding of its own strategy, but also delineating its enemies in class terms, for without this the
pragmatic eclecticism threatened merely to refresh and renew the
existing social order. Providing the enemy used the same language of
moral concern — and sometimes even the same populist rhetoric — it was
difficult to distinguish friend from foe. For instance, when the Club of
Rome — composed of those same industrialists and scientific elite who
had been in charge of the production/pollution process — announced
their collective concern with the finite earth, then, lacking a class perspective, the movement seemed only to carp ungenerously at a deathbed
repentance. The invocation of 'the scientific community', like that of
'the national interest', sought to bind strata with an antagonistic
relationship into one ideological whole.

The magnitude of the theoretical tasks confronting the movement –
the need for a political economy of science in contemporary capitalism,
its changing mode of production, the proletarianisation of scientific
workers, the question of natural science as a generator of ideology, and
of the ideology of science with its devaluation of all non-'scientific'
knowledge, its elitism and the subtleties of its particular form of sexism
and racism – all these needed definition and welding together theoretically. We had to achieve these tasks in the knowledge of the past
history of theory and practice on the question of science in the revolutionary Marxist movement – and, in particular, the experience of the
Soviet Union and China. Such an agenda was daunting for each of us
individually – yet we all believed that to tackle it on the basis of our
own separate experiences in different capitalist countries was imperative, and that collectively we could make a start. Geographical distance
between us has meant that this has not been a fully collective programme in the sense that all its authors have participated in the writing
of all sections, but rather that each has taken a particular section of the
agenda and developed an analysis within a general shared framework,
whilst in a few cases we have used material which was not written
specifically for these books but seemed clearly in accord with their
over-all theoretical position. By common consent, all royalties from the
publication of the collection, in the several languages in which they are
to appear, will go towards the development of scientific and technological education and reconstruction in Vietnam, by way of the Institute
for Science and Technology in what was once Saigon, and is now Ho
Chi Minh City, part of our recognition of the imperishable role that the
struggle and sacrifice of the Vietnamese people has played in the theory
and practice of revolution and of the transformation and recreation of
human society.

*The Political Economy of Science*
The collection of essays have been organised into two volumes, with the
common theme of ideology in/of the natural sciences. Whilst the two
books are separate entities they reflect certain common concerns and
are interrelated by a logical thread which this Introduction traces.
The starting point has been an attempt to transcend both our own
particular political pasts, and that of the revolutionary movement,
which for too long has seemed to be polarised between, on the one
hand, 'orthodox Marxism' with a rigid belief in the objectivity of the

natural sciences as a model to which Marxism, as scientific socialism, aspired, and on the other, an anarchism which has seen scientific rationality itself as part of the enemy. In order to recreate a revolutionary critique of the actual social functions of science as they exist in today's capitalist and state socialist societies, it is necessary to understand the origins and the limitations of this 'orthodox Marxist' view of science, which regards itself as operating in a tradition which stretches from the most recent pronouncements of the Soviet Academy of Sciences back through Stalin and Lenin to Engels, and hence Marx himself. We therefore begin, in the first chapter of *The Political Economy of Science*, by returning directly to what Marx and his close collaborator Engels themselves wrote about science, and in doing so rediscover in Marx those compelling theoretical insights which, however briefly and schematically they are presented, lie at the core of every one of those questions of theory and practice which are the concern of today's movement.

The second chapter of *The Political Economy of Science* moves directly forward from Marx and Engels to the issues of the 1970s, with which the whole of the rest of the book is concerned. In 'The Incorporation of Science', we ask what features characterise the present social function of science in Western capitalist societies and the Soviet Union. We argue that, today, science has two major functions, as part of the systems of production and of social control. Especially since the Second World War, science has itself become industrialised and enmeshed in the machinery of state. We examine two myths, the liberal academic myth of the autonomy of science and the 'orthodox Marxist' belief in the inevitable contradiction between science and capitalism, and show that neither accounts for the actual development of science and 'science policy' — the management of science — as it has occurred in Britain or the United States. Faced with capitalism's fusion of science and oppression, and the conspicuous failure of the Soviet Union to avoid the same development, the 'Frankfurt School', typified by such writers as Habermas, has claimed that scientific rationality is *inevitably* oppressive and has abandoned that optimism with which Marxists had maintained the automatically progressive nature of science. The question is whether capitalist science represents an unavoidable and fatal attempt at the domination of nature, or whether it can be confronted as a 'paper tiger', to make way for a genuine science for the people.

The next four chapters discuss in greater depth the questions, raised in Chapter 2, of the role of science in production and the consequences for scientific workers. This issue raises fundamental questions for Marxists both at the theoretical level and in terms of political and organisational strategy. In the first place, where does science fall within the Marxist categories of 'base' and 'superstructure'? Is it part of the productive process? This is not an abstract question, for if it is purely superstructural, then scientists, whatever the contradictions within their role, cannot be regarded as workers, but primarily as within or associated with the ruling class, either by assisting in the structural maintenance of the capitalist apparatus, like lawyers or accountants, or as transmitters of its ideological values, like teachers or journalists. That is, they will in general find that the contradictions of capitalist society do not oppress them but serve to protect their privileges and position. On the other hand, if science is part of the productive process, 'scientists' are really scientific *workers* who sell their labour to the capitalist in parallel with other workers; like other workers, they become alienated from their creations, from the products of their labour – in a word, they are proletarians, and as such form part of the potential revolutionary forces within society.

This issue has long been a source of debate and discussion because upon it hangs the question of whether, politically, scientists are to be seen as friend or foe. This is particularly important in the present period of the incorporation of science, and the answers given by Marxists in earlier periods may no longer be appropriate today. These chapters argue, essentially, that science spans *both* base and superstructure; it has both a productive and an ideological role, the understanding of which is confused by reference to 'the scientific community' as an undifferentiated whole. In fact, this 'community' is divided into, on the one part, the majority of alienated, proletarianised *scientific workers*, and, on the other, the tiny majority of the elite carriers of bourgeois ideology, the *scientists*.

Chapter 3 is by a group of physicists and mathematicians, Giovanni Ciccotti, Marcello Cini and Michelangelo De Maria, associated with the *Manifesto* group in Italy. They approach the question of the role of science as a productive force from the perspective of Marx's theory of value. Today, they conclude, the role of applied science and technology can be seen as the production of information as a commodity, to be sold on the market just as are material commodities. The relation of scientific workers to their product is therefore comparable to that of manual workers; they are alienated from it. Science as commodity production is thus the dominant mode, which serves as a model for the style of work even in fields which are not directly concerned with the production of information for sale, such 'pure' sciences as high-energy physics or biology. These fields have a dual role, generating an information 'base' on which the information-commodity market can rest, and serving as test-beds for the checking of advanced technology.

Chapters 4 and 5 take up the consequences of this role of science as a productive force for scientific workers themselves. André Gorz, the editor of *Temps Modernes*, asks: what are the implications of describing scientific workers as proletarianised? Science is still a privileged, elite activity: in industry scientific methods may be used by some categories of workers (production engineers for example) to oppress others by means of speed-ups and other forms of technological rationalisation; none the less, the fragmentation of scientific knowledge, and its ideological values, has come to make intellectuals increasingly the victims rather than the beneficiaries of the class system. The way forward lies in ridding expertise of its class nature, of breaking the barrier between expert and non-expert.

To a large extent, Mike Cooley shares Gorz's preoccupations, but brings to them the perspective of the shop-floor struggles which his own designers' and draughtsmen's union (TASS, a section of the Amalgamated Union of Engineering Workers, AUEW) has been involved in. Cooley shows how the increasing cost and rapidity of obsolescence of fixed capital impose increasing demands on both manual and intellectual workers in industry, with speed-ups, shift work, fragmentation of skills and dehumanisation. This proletarianisation began, as Gorz points out, in the chemical industry in the nineteenth century, but has now spread to designers and draughtsmen, architects, computer programmers and mathematicians in industry. However, as Cooley shows, a capitalism based on very complex, very expensive technology, develops the weaknesses of its own strengths. It is these points of vulnerability which proletarianised scientific workers, side by side with their manual worker comrades, must learn to probe and enlarge if the system is to be shattered and social transformation to occur.

The remaining chapters of *The Political Economy of Science* are concerned with a distinct theme, whose roots, as we show in Chapter 1, derive from Marx's and Engels' own writings, but which has burgeoned into major significance in recent years. This is the theme of the struggle between ideology and science within the natural sciences themselves. The analysis of this struggle is no easy task. Ideology is of its nature mystifying. Where the sharpness of the contradictions within the capitalist mode of production continually force themselves into the consciousness of the worker, the very role of ideology is to obscure these contradictions and diminish the level of consciousness. Hence, whilst the superstructural battle and that in the work-place are part of the same conflict — indeed, they continuously interact — the dominant class pretends that there is no ideology, and so no grounds for battle: that science has once and for all driven out all ideology. In the second place, because of the abortive nature of the Soviet cultural revolution and the experience of Lysenkoism (discussed in Chapter 2 of *The Political Economy of Science* and Chapter 2 of *The Radicalisation of Science*), the continuity of the critique of ideology has been ruptured. Marxists are faced not only with the problem of starting afresh from the moment of rupture, but also with the analysis of the rupture itself. For many years, orthodox Marxism in its preoccupation with the objective world laid to one side complex questions of the superstructure, arguing for the most part that it was determined by the economic base; natural science, while belonging to both, was above ideology.

Yet battles in the superstructure are not some revolutionary luxury item which can be dealt with after the workers have destroyed capitalism, but are intrinsic to the political struggle itself. No one writing in these books has gone out to look for 'ideology in astrophysics', 'ideology in inorganic chemistry', in cell biology, biochemistry, and so on in the way which it seems Marxist scientists did in the 1930s, clutching their *Dialectics of Nature* and searching for thesis, antithesis and synthesis in the particular bit of the natural world they worked in. Instead, work on science's role in perpetuating racism, exposing the implications of reproduction science for women, or the nature of the politics of ecology, has been written as part of an on-going struggle, not as an item of an academic agenda. For this reason these chapters do not represent an even spread over the natural sciences. So long as most of the current struggles relate to the biological sciences, then it is right that we work in this area. (It is not however the case that the cultural analysis in some sense 'follows' the existence of struggle at the point of production, nor is it a question of awarding prizes for priority in discovering racism to the Mansfield hosiery workers or those working on scientific racism, but rather that each should see the other as necessary.)

Chapters 6 and 7 of *The Political Economy of Science* interlock, in that the second, on scientific racism, is a special case of the critique of ideology in the neurobiological sciences contained in the first. Both chapters argue that many of the theories and linked technologies of neurobiology, from drug therapy through behaviour modification to IQ testing, are fundamentally biologistic. Biologism takes one part of the explanation of the human condition, excludes all other considerations, and announces that it has *the* explanation for aggression and altruism, war and class struggle, love and hate. Attempting to change the human condition is then presented as an absurd opposition to both our natural selves and the natural world. The everyday possibility and actuality that men and women have continuously changed their situations in the course of history is methodologically and philosophically excluded. Biologism, for all its apparent scientificity, is thus mere ideology, the legitimation of the *status quo*. It is a method not of explaining people, but explaining them away as 'nothing but' assemblages of molecules, larger rats, naked apes or hairy computers. In biologism, reductionism, which was originally simply a powerful tool for examining specific problems under rigorously defined conditions, becomes saturated with ideology. Reductionism is thus part of the ideology *of* science, and in so far as the theories serve specific dominant classes, also legitimises and obscures ideology *within* science. The particular importance of biologism derives from the nature of the fight in which the bourgeois state must presently engage to protect itself. Where in the past its military effort was primarily against other nation-states or directed towards securing new colonies, with internal control a related but subsidiary question, since the growth of revolutionary guerilla movements, the main enemy is within. Faced with this internal enemy, methods of social control become of paramount importance to capitalism; biologism with its ideological justification and its techniques of manipulating and controlling people comes to the rescue.

Chapter 8 of *The Political Economy of Science*, while still concerned with biology, sets out to analyse the ingrained sexism of current developments in reproduction technology, from genetic engineering to hormone time capsules. This characterisation of science is opposed to that of the radical feminists such as Shulamith Firestone who see technology as essentially neutral and therefore capable of generating a 'technological fix' for the reproductive role of women. By contrast, the chapter argues the need to link the class and the women's struggle in the pursuit of human liberation, where science would serve the goal of nature humanised, and 'the long struggle from nature to a truly human culture' would be advanced.

The final chapter, by Hans Magnus Enzensberger, West German poet and political activist, is a critique of political ecology. In it, Enzensberger is concerned with two tasks. One is to expose the ideological role played by the prophets of the ecology movement as it has mushroomed since the late 1960s, people like the Ehrlichs, Forrester and Meadows, the MIT modellers of 'the limits to growth' and the 'Club of Rome'. Enzensberger lays bare the links between the 'ecology movement' and imperialism, and shows that in their frequent apocalyptic pronouncements, the doomsters are playing a deeply ideological role. The second point is that the concern over pollution or global destruction cannot be dismissed as pure ideology or merely a consequence of capitalism that the transition to socialism will automatically resolve, as some Marxist groups tend to argue; this itself becomes an ideology which ignores the real material base for much of the present concern. The ecological hazards are not to be dismissed as trivial, and even after the destruction of capitalism they will remain major problems. 'Socialism, which was once a promise of liberation, has become a question of survival. If the ecological equilibrium is broken, then the rule of freedom will be further off than ever.'

## The Radicalisation of Science

Whilst *The Political Economy of Science* is concerned primarily with the critique of existing capitalist science, much of the discussion in *The Radicalisation of Science* deals with attempts at its transformation. The first chapter of the book, originally written for the 1972 issue of *The Socialist Register* and subsequently reprinted in *Science for People*, gives the book its title. It represented the gathering together of our personal experiences within the scientists' movement at that time, an attempt to describe the origins, the brief history and perspectives for action of the movement. Even though our understanding of certain of the issues has sharpened in the intervening period, we decided to reprint it as it stands, both because it has served to fuel a necessary debate within the movement in the last few years, and because it represented the original programmatic guide for the present collection. However, we have updated it, and added a postscript from the vantage point of 1976. Chapter 2 takes up a topic which no discussion of the relationships between Marxism and the natural sciences can avoid. This is the Lysenko 'affair', 'problem', 'scandal' — as it has variously been described. Coming at a crucial time in the development both of the Soviet Union and of the attempts by Marxist scientists in the West to grapple with the problem of the relationship between science and social structures, it seemed to provide the acid test of the possibilities of a socialist science. The consequences of the debate were disastrous — concretely for the geneticists who lost their lives in Stalin's camps, for the development of Soviet genetics (and, less certainly, Soviet agriculture) and theoretically for the very idea of a socialist science. The period following 1948, the high point of Lysenkoism, marked a retreat in the Soviet Union to a 'neutral ideology of science, and, in the West, a turning away of many scientists from the orthodox communist parties and even from Marxism itself; they were 'forced to choose between their science and their political convictions'. As the period of Lysenkoism retreats, so it gains a mythology, and even Marxists have shied away from attempting to peel off these mythical accretions so as to subject the episode itself to rigorous Marxist analysis. Yet it is essential that we understand what happened, it only to help avoid a repetition of old mistakes. As Richard Lewontin and Richard Levins make clear, it is no good merely to see the episode as an example of the workings out of the 'cult of the personality', or a dreadful warning of the consequences of mixing biology and politics — nor yet as the high point of Soviet science before its retreat with the rise of revisionism. Rather, we must seek its roots in the objective conditions of Soviet agriculture and society, and understand it as an aspect of the tentative and inadequately articulated attempts within the Soviet Union of the 1930s and 1940s to achieve a cultural revolution — but one monstrously distorted by its imposition 'from above' by a mixture of administrative fiat and terror, rather than 'from below' by a creative social and political upsurge amongst the people themselves.

What was Lysenkoism most directly about, and what were its claims? As mathematical biologists, whose own research relates directly to the substance of the Lysenkoist claims, and as themselves politically engaged within the Science for the People movement in the United States (both refused membership of the US National Academy of Sciences on the grounds of its involvement with the Department of Defense and its perpetuation of the hierarchical, elite structure of American science) Lewontin and Levins are well placed to make the assessment. They begin by assessing the present significance and interest of the Lysenkoist controversy. They then briefly summarise the philosophical and scientific claims of Lysenkoism itself: what were Lysenko's views on heredity and its relationship with the environment? (It might be helpful to those unfamiliar with the details to compare this discussion with that in Chapter 8 of *The Political Economy of Science*, where some of the same issues are discussed in relation to the IQ debate.)

Lysenko's views are contrasted with some of the almost mystical concepts which many classical geneticists of the Weismann school at the time held about the gene and its relationship to the environment. Then, in a crucial section of the argument, they discuss the objective conditions creating Lysenkoism: the weakness of Russian agriculture and its climatic problems, and the implications that these latter had for the interpretation of experiments and the use of statistics. The weaknesses of existing genetic theory, and its ideological role and links with philosophical reductionism and racism, are analysed.

Other vital factors were the reaction of the Russian peasantry to collectivisation, and the elite, bourgeois structure of Russian science which still remained the case even twenty years after the 1917 revolution. It is this feature – the challenge to the bourgeois expert – which represented that part of Lysenkoism which can be seen today, with the hindsight provided by the Chinese experience, as the attempt at cultural revolution.

Lewontin and Levins conclude by asking: Can there be a Marxist science? The answers they give, in terms of what the dialectical method can and should mean in science, may serve, in their emphasis on the unity of structure and process, the wholeness of things and the interpenetration of an object and its surroundings, as a key and summary statement of the major themes of both books.

The next two chapters of *The Radicalisation of Science* are, concerned with the nature of the institution(s) of science as they have developed under contemporary capitalism, and particularly its sexist character. Monique Couture-Cherki, a solid-state physicist from Paris, and Liliane Stéhelin, a sociologist of science from Strasbourg, raise the question of sexism. Couture-Cherki points to the systematic exclusion of women from the higher ranks of science, their concentration in subordinate positions, and the powerful ideological pressures which are exerted to systematically exclude women from scientific achievement. Amongst these, the most powerful are the ideology of the family and the persistent attribution to women of more 'docile', 'feminine' characteristics, 'not appropriate to high scientific achievement', and so on. But can these be overcome? Liliane Stéhelin takes this question as her starting point. For her, the present forms of science are fundamentally interlocked with sexist, male ideology. In order to succeed in science, a woman is required to submerge – overcome – her feminine character and become an honorary male. To do this is the ultimate trap. Indeed, we can expect, at least in periods of labour shortage and capitalist expansion, to see a steady effort made to eliminate the obvious barriers to women's progress in science, the provision of creches and better maternity arrangements, more efforts at 'equal opportunity' appointments, and so forth – if only because women represent a reserve of productive forces.

Yet the production code of science, its ideology, will remain fundamentally masculine; forced to compete within it, women will either succeed by denying their femaleness, or fail, confirming their inferiority. The task, therefore, is the attack on and subversion of the masculine code itself, which raises the question of whether there is indeed a feminine science as an alternative to masculine science in the same way as there is a socialist as opposed to a bourgeois science. This question leads Stéhelin into a consideration of the social and psychoanalytic view of women and into the question of the resynthesis of Marxism and psychoanalysis which has been a major concern of French Marxism in recent years. Can the masculine code of science be overcome? If so, she concludes, there is 'the promise that one day other women (with other men?) will be able to open the way for a new science'.

It is against this background that it becomes possible to raise the question of just what can be learned from the Chinese experience. Despite the greater accessibility of China, and the enthusiasm for what are seen as the lessons of the cultural revolution, an adequate account of what has been and is being achieved in China must start from an understanding of the particular circumstances of China's own social and economic development, rather than from timeless universals. Joseph Needham's chapter was originally given, in 1975, as a lecture in Montreal, and its lecture form is preserved here. In it, he first describes his own history and that of the *Science and Civilisation in China* project (Cambridge University Press, 1954 onwards), and then sets out to counterpose the historical development of Chinese science with the contradictions of science and the anti-science movement in Western capitalism as typified by, for instance, Theodore Roszak. Needham argues that the anti-science movement has emerged in the West in response both to the social function of science under capitalism and the claim that science represents the only valid way of understanding and apprehending the universe – an aspect of the scientistic ideology of science with its overriding aim of the domination of nature. By contrast, he shows, the Chinese have historically never had such a scientistic approach nor fallen prey to reductionism. This is not to say, Needham emphasises, that the practice of science in today's China has nothing in

common with that under capitalism, but it is a practice reflective of a
dialectical conception of the interrelations of nature and humanity, and
of a science done for and with the participation of the people as a
whole. Needham's analysis is couched in characteristically more ethical
and religious language than is familiar to many activists in the radical
movement today; a language from within the tradition of English Chris-
tian communism, its moral passion echoing that of Digger Winstanley.

Chapter 6 is derived from an article in *The Black Scholar*, 'Science,
Technology and Black Liberation', by Sam Anderson, a New York
mathematician. In it, Anderson briefly outlines some reasons for the
technological underdevelopment of Africa by European colonialism and
the role of science in the emergence of capitalism, leading to the present
situation in which, for the Third World countries, science has the two
aspects of 'liberation' and 'exploitation'. The position of the black
scientist in the United States (or Western Europe) has much in common
with that of the woman scientist discussed by Couture-Cherki and
Stéhelin — forced into an alien, bourgeoisified role. To combat this,
and to contribute needed scientific and technological skills for the
movement, Anderson calls for black scientists to organise.

The final chapter, by Jean-Marc Lévy-Leblond, theoretical physicist
and one of the collective producing the radical science magazine
*Impascience*, spans the themes of both ideologies, *of* and *in*. Because
modern physics is a discipline founded at the birth of capitalism, it is,
in certain important respects, the model to which all science aspires.
Although its theories may have little ideological significance in themselves, physics as a social and cognitive institution is saturated with
capitalist ideology, and the ideology of physics as a science becomes the
dominant theme of Lévy-Leblond's chapter. To mathematise, to forma-
lise, becomes the hallmark of the mature, hard science against the
immature, soft science (the masculinity/femininity — superior/inferior
metaphor is not lost). Nor is this only an issue in the natural sciences, as
physics becomes the model for all human knowledge, and what cannot
be encompassed by its mode of rationality is illegitimate.

Physics is thus at the heart of the ideology of expertise: the claim
that, to be a physicist, particularly a theoretical physicist, gives an
individual *as of right* the power and knowledge to speak with competence in almost any area.

Within physics, social practice is deeply hierarchical between scientist and student or technician — symbolised by science's reward system,
at the peak of which come the Nobel Prizes. The Laureate, in fact a

narrow specialist, becomes transmuted by social alchemy into one of Plato's Men of Gold, to whom all humanity must defer. Another aspect of the hierarchy though is the divorce between theory (high prestige) and practice (low prestige), epitomised by the elite nature of theoretical physics and the lower status of the experimental science of engineering. Lower still, yet equally hierarchised, comes teaching. This divorce affects the development of the subject of physics and, at the same time, lays it open to the type of ideological exploitation discussed in relation to biology in other chapters. The divorce from practice means that physicists are increasingly concerned with an artifical world of their own construction, outside the experience of common problems which physics used to be concerned to explain. The solution for these problems will be the solution for science as a whole.

*****

The themes of the chapters in these two books reflect a common agenda, an agenda shared with many of the activists in the radical science movement who have been discussing and working out these issues in practice over the last few years. At an earlier stage, many of the chapters have formed part of, and been improved by, this discussion. By collecting and developing the arguments on paper, we believe that the theory and practice of the movement will be advanced. Nevertheless, it is important not to forget differences. These reflect the fact that we belong to a social movement with diffuse aims and not to a single party with a clear line and agreed priorities. What we hold in common is a desire to work towards a new society where a new science and technology can serve the interests of all the people.

# 1

# The Problematic Inheritance: Marx and Engels on the Natural Sciences

## Hilary Rose and Steven Rose

This chapter returns to the writings of Marx and Engels on the nature of scientific knowledge and the relationships between science and capitalism. Marx himself did not specifically treat these questions in isolation from his general analysis of consciousness, ideology and modes of production (nor indeed should we expect him to do so), as for Marx and Marxism these themes are but part of a seamless web in which analysis of the social functions of science and its value as a mode of knowledge of the material world are integrated as a consistent strand through all the writings, from the early period of the *Economic and Philosophical Manuscripts* and *The German Ideology* through to *Capital* itself. Our method in the first part of this chapter is so far as possible to allow it to speak in Marx's own words, with the minimum of commentary, although there are areas where, as in the discussion of alienation and of ideology, it is necessary to develop the account inside the practice of science as a special case of Marx's more general analysis. It was left to Engels to develop the common themes, which he and Marx had discussed, into a more coherent account of science, and the main legacy of Engels' writing in this area is *Dialectics of Nature*, which was first published in 1925 some thirty years after his death. Engels' analysis in *Dialectics* has raised serious problems of both theory and practice for Marxists, which are reflected at all levels in today's struggles, and we therefore treat our discussion of it in the late sections of the chapter rather differently, so as to draw out the implications of Engels' position and the subsequent debate it aroused. We do not pursue the long history of this debate — although it would be well worth doing — through Lenin to Lukacs and Althusser. Instead we focus here on the legacy of Engels, for we believe that it is in this problematic contribution that the subsequent difficulties have their origins. We begin, however, with Marx's account of the nature of human knowledge and the knowledge of natural history.


## THE UNITY OF THE HISTORY OF NATURE AND OF HUMANITY

For Marx the history of nature and the history of humanity are unintelligible if divorced from one another. Nature can only be seen and understood in relation to humanity. So that while it is possible to envisage a natural world prior to the existence of man,* or indeed after man has ceased to exist, for man these have no meaning: 'But nature too, taken abstractly for itself — nature fixed in isolation from man — is nothing for man.'[1] In the same way the history of humanity can only take place within an objectively real world — a world of nature which is itself continuously changing through human actions.

> The first premise of all human history is, of course, the existence of living human individuals. Thus the first fact to be established is the physical organisation of these individuals and their consequent relation to the rest of nature. . . . The writing of history must always set out from these natural bases and their modification in the course of history through the action of man.[2]

> the sensuous world . . . is not a thing given direct from all eternity, ever the same, but the product of industry and of the state of society; and indeed in the sense that it is an historical product, the result of the activity of a whole succession of generations. . . . Of course, in all this the priority of external nature remains unassailed . . . but this differentiation has meaning only in so far as man is considered to be distinct from nature. For that matter, nature, the nature that preceded human history . . . today no longer exists anywhere.[3]

Thus the tasks of actively comprehending the natural and the social

---

*We are unhappily aware that even when 'man' and 'he' are intended as generic abstractions for humanity, in the alchemy of reading they become transmuted into images of masculine men, and the half that are women have been dissolved from history once again; we use the term in this chapter for consistency with the quotations from Marx.

world are in the last analysis not two separate intellectual tasks but one.

> We know only a single science, the science of history. History can be contemplated from two sides, it can be divided into the history of nature, and the history of mankind. However, the two sides are not to be divided off; as long as men exist the history of nature and the history of men are mutually conditioned.[4]

What distinguishes human history from the history of the rest of nature, including that of other animals, is that man has learnt to produce the means of his own existence. And hence men produce their own material life. This production occurs by acting upon nature, and it is through this practice that the social and natural worlds are changed. The very fact that human beings learn to be more productive, to objectify themselves, to create more material objects, makes possible, and actual, new ways of living.

This first practice springs from man's conflict with nature, including other men as part of nature, in order to survive. (Not for nothing did Marx want to dedicate *Capital* to Darwin.) Knowledge comes from this conflict with, and action upon, nature. Natural science is the form of knowledge of the natural world which developed in the specific historical context of capitalism.

> Labour is, in the first place, a process in which both man and nature participate, and in which man of his own accord starts, regulates and controls the material reactions between himself and nature. He opposes himself to nature as one of her own forces, setting in motion arms and legs, head and hands, the material forces of his body, in order to appropriate nature's productions in a form adapted to his own wants. By thus acting on the external world and changing it, he at the same time changes his own nature. . . . Nature becomes one of the organs of his activity.[5]

Thus the central concept in Marx's theory of knowledge is that of practice, for it is through practice that we both know and change the world. This has been put nowhere more clearly than by Mao Tse Tung:

> If you want knowledge you must take part in changing reality. If you want to know the taste of a pear you must change the pear by eating it yourself. If you want to know the structure and properties of an atom, you must make physical and chemical experiments to change the state of an atom. If you want to know the theory and methods of revolution, you must take part in revolution.[6]

The difference between Marx's theory of knowledge, with its integration of theory and practice, and that of bourgeois culture, is that the latter emphasises the separation of knowing and doing, theory and practice, science and technology, natural science and social science. In opposition to this Marx asserts the unity of historical method, and Mao the continuity between understanding atoms and understanding revolution. Because the history of humanity is the history of nature developing into humanity, 'Natural science will in time incorporate into itself the science of man just as the science of man will incorporate into itself natural science: there will be *one* science.'[7]

## SCIENCE IN CAPITALIST PRODUCTION

Because scientific knowledge is both achieved in the practice of transforming nature and itself becomes an agent in the transformation of human nature, its development goes hand-in-hand with changes in the mode of production, first with handicraft and then with industry. As pre capitalist social forms were transformed into capitalist ones, so empirical folkloric knowledge was transformed into the first phase of modern science. Science, like capitalism, was a civilising force — within limits. To give two examples, the Galilean revolution in cosmology destroyed the Ptolemaic model of an earth-centred universe neatly ordered by God. Later, Darwin was to make God redundant from the creation of life and of humanity. Science thus appeared as critical knowledge, liberating humanity from the bondage of superstition, a superstition which, elaborated into the thought system of religion, had acted as the key ideological prop of the outgoing social order. The system of production founded on capitalism requires this continuous innovation in all spheres of life: the creation of new objects, new ideas, new technologies and new social forms. It requires, Marx tells us, 'the development of the natural sciences to their highest point.'[8]

He goes on to indicate how under these circumstances science becomes a direct force of production:

> Nature builds no machines, no locomotives, railways, electric telegrams, self-acting mules, etc. These are the products of human industry; natural material transformed into organs of the human will over nature, or of human participation in nature. *They are organs of the human brain, created by the human hand*; the power of *knowledge objectified*. The development of fixed capital indicates to what

degree social knowledge has become *a direct force of production* and to what degree, hence the conditions of the process of social life itself have come under the control of the general intellect and been transformed in accordance with it.[9]

At the same time capitalism contains within itself a contradiction which both 'limits the civilising process', and, by limiting it, builds up an irresistible force for capitalism's own destruction. This contradiction is, of course, that between labour and capital.

But what of science? Does it stay in the Galilean/Darwinian manner and continue as a force for liberation, or does it share in the general contradiction? Marx is unambiguous on this. Under capitalism, nature becomes denatured, humanity dehumanised; and science is integral to both processes.

> For the first time nature becomes purely an object for mankind, purely a matter of utility; ceases to be recognised as a power for itself; and the theoretical discovery of autonomous laws appears merely as a ruse so as to subjugate it under human needs, whether as an object of consumption or as a means of production.[10]

As material wealth grows through the capitalist mode of production, the worker finds himself increasingly devalued, alienated from what he produces, from himself, and from his species being. 'The *devaluation* of the world of men', writes Marx, 'is in direct proportion to the *increasing value* of the world of things'. The relationship of the worker to production simultaneously produces both the immense power and wealth of labour and powerlessness and privation for the worker. The things he produces are not his; he may build a luxury flat, but he himself is ill-housed or even homeless, so that what he creates with his own hands appears as a hostile alien object, belonging to others, not to him. In addition the worker is alienated in the very *act of production*; he does not work because he feels the *need* to work, but in order merely to exist. Work thus becomes a continuous sacrifice of the individual self. Because, for Marx, production — that is the capacity for man to objectify himself, to create objects — is the distinctive human activity, estrangement from this is to be estranged from the human species itself.

The point is, as we said earlier, that science plays an integral part in the alienation of labour:

> natural science has invaded and transformed human life all the more *practically* through the medium of industry; and has prepared

human emancipation, although its immediate effect had to be the furthering of the dehumanisation of man. *Industry* is the *actual*, historical relationship of nature, and therefore of natural science, to man.[11]

This estrangement through natural science is revealed by the specific technological forms which industry employs. Not only is science in a general sense incorporated into the production process, thus intensifying human alienation (although simultaneously preparing the grounds for its dissolution), but as the productive capacity of industry develops, science becomes in itself a direct force of production. A force which as it grows more powerful bears as its correlate the weakening, the intellectual and physical deskilling of the worker, who in the scenario Marx outlines (and this in 1848) becomes a mere appendage to the machine.

> Intelligence in production expands in one direction because it vanishes in many others. What is lost by the detail labourers is concentrated in the capital that employs them. It is a result of the division of labour . . . begins in simple co-operation . . . developed in manufacture which cuts down the labourer into a detail labourer. It is completed in modern industry, which makes science a productive force, distinct from labour and presses it into the service of capital.[12]

> the *direct* relationship between the *worker* (labour) and production . . . produces intelligence — but for the worker, stupidity, cretinism.[13]

SCIENCE AS ALIENATED LABOUR

Although Marx and Engels do not discuss at any length the extension of the machine into the production of modern science, and the shift therefore, within science, from handicraft manufacture to industrial manufacture, science is clearly seen as a branch of production subsumable under the general laws of production. Thus, where in the nineteenth century, individual scientists worked singlehandedly (or aided by their single technician/apprentice), scientific workers are now for the most part grouped around large-scale equipment housed in great laboratories. What the Spinning Jenny did to create factory workers, the particle accelerator does for the scientific workers: 'Religion, family, state, law, morality, science, art, etc., are only particular modes of production and fall under its general law.'[14]

Because the socialisation of scientific production has followed the

socialisation of general production, the view of scientific work as alienated labour was only to become fully visible after both Marx and Engels'
deaths. As the 'general intelligence' of science grows, the individual
scientific worker is deskilled both in terms of his intelligence and his
manual skills. Further, as with all alienated labour, the scientific
worker's product itself serves to oppress him — and his fellow workers —
a process apparent even in the mid-nineteenth century, as Marx notes
in *Capital*: 'It would be possible to write quite a history of the inventions, made since 1830, for the sole purpose of supplying capital with
weapons against the revolt of the working class.'[15] But this thesis, for
which the theoretical framework exists within Marx, is explored again
and again by writers within these volumes.


SCIENCE AS IDEOLOGY

Marx and Engels repeatedly emphasised the divorce which occurs in
bourgeois culture between knowledge of the natural world and knowledge of the social world. Knowledge of the social separated from the
natural becomes idealism, knowledge of the natural separated from the
social becomes mechanical materialism — itself a form of idealism.
Within capitalism each becomes elevated into a world view which can
serve only to prevent the realisation of true knowledge: idealism by
giving primacy to ideas, to human consciousness separated from the
material circumstances which create it; mechanical materialism by
giving primacy to nature separated from humanity's actions upon it.
These world views constitute the dominant ideas within capitalist
society; they are held and promulgated by, and on behalf of, the ruling
class. As ideologies they serve to obscure the actuality of the world,
thus helping prevent the possibility of changing it:

> The ideas of the ruling class are in every epoch the ruling ideas; i.e.
> the class which is the ruling material force of society, is, at the same
> time, its ruling intellectual force. The class which has the means of
> material production at its disposal, has control at the same time over
> the means of mental production, so that thereby, generally speaking,
> the ideas of those who lack the means of mental production are
> subject to it. The ruling ideas are nothing more than the ideal expres-
> sion of the dominant material relationships.[16]

Marxism counterposes ideology, the 'ruling ideas', with science, conceived of as providing the means of understanding, interpreting and

changing the natural, human and social worlds. It has always been apparent that the social sciences are susceptible to ideological penetration, or may even be totally captured, so that official social science becomes in its turn an ideological weapon.

Althusser reminds us of the constant struggle to free science from ideology, contrasting the open, question-posing character of science with the closed dogmatic character of ideology; in *For Marx* he writes:

> There is no perfectly transparent science, which, throughout its history as a science will always be preserved . . . from . . . the ideologies that besiege it; we know that a 'pure' science only exists on condition that it continually frees itself from the ideology which occupies it, haunts it or lies in wait for it. The inevitable price of this purification and liberation is a continued struggle against ideology itself.[17]

The possibility that the natural sciences might be similarly occupied does not seem to have been explored by Marx and Engels themselves, partly perhaps because of Engels' optimistic belief that as the only approach to true science was through dialectical materialism, working scientists would be forced by the nature of their subject to become dialecticians. However, in a comment on Darwin's *Origin of Species* made in a letter to Engels, Marx noted the way in which Darwin appeared to have looked at the biological world, and found mirrored therein the class structure of Victorian capitalism — a mirror which the Social Darwinists were also to use for their own ideological purposes. But the subsequent exploration of this important insight was to become lost in the crude application of Lenin's copy—theory of reality in his *Materialism and Empirio-criticism*.

As capitalism has developed, mechanical materialism — described variously more recently as 'scientism', 'positivism', 'hard-nosed objectivity' — has become the dominant ideology. Science as ideology has extended the reductionist methods of natural science over the human sciences as well, turning human subjects into objects (that is, making the potentially active, passive) — at its logical extreme reducing human beings to 'nothing but' the abstract categories of biological and chemical laws. All knowledge except that which is legitimised by this mechanical materialism is denied. The consequence, as the 'scientists' are the producers of mechanical materialism, is that science becomes an ideology and scientists the ideologists. How does this work? As the material world controls the limits of an interpretation of the scientist in his *own* work, the answer lies, as Marx and Engels saw, *outside* the precise research area, where the scientist, freed from such constraints, talks (typically in the name of science) pure ideology. In the name of science, invoking neutrality, technique and expertise, the scientist supports the ruling strata:

> The weak points in the abstract materialism of natural science, a materialism that excludes history and its process, are at once evident from the abstract and ideological conceptions of its spokesmen, whenever they venture beyond the bounds of their own speciality.[18]

Historically it has proved hard to see 'science as ideology' because of the important role of science in driving out the religious ideology of pre-capitalist society. Thus natural science, like capitalism itself, has had a progressive character, criticising and destroying feudal ideas and social formations. Like capitalism it is limited, and in its turn becomes more oppressive than liberatory.

## ENGELS AND THE DIALECTICS OF NATURE

Marx had a continuing interest in theories and developments within the natural sciences; together with William Liebknecht, he attended T. H. Huxley's lectures and followed the progress of chemistry, physics and biology. As Liebknecht recalled in his graveside *Reminiscences of Marx*, 'when Darwin drew the conclusions of his investigations and made them public, for months we talked of nothing else but Darwin and the revolutionising power of his scientific achievements'.[19]

None the less, Marx's writing is primarily concerned with science in relationship to industry and as a force within the capitalist mode of production, so that it was left to Engels to carry out, after Marx's death, the work on natural science they had planned together. Engels' work is set out in three main texts, two polemical books published within his lifetime: *Anti-Dühring* (1885) and *Ludwig Feuerbach and the End of Classical German Philosophy* (1888); and a synthetic but fragmentary text first published only after his death, *Dialectics of Nature* (1925). Because both of the two sides of the long debate within Marxism over the status of natural science can be seen to have their roots within — or in reaction against — Engels' writing, this section on Engels is treated rather differently from that on Marx, as while Marx is unambiguous if fragmentary on science, Engels' work is the main inheritance. While much of his work is theoretically co-terminous with that of Marx, almost as soon as the *Dialectics* was published, a major criticism was levelled at it by Lukacs, who argued that by transferring the dialectic from between man and man, or man and nature, and locating it in nature itself, Engels had in an important respect departed from a materialist philosophy of history. Dialectics in nature took on an ontological status, a fundamental property of the universe: 'The misunderstandings that arise from Engels' dialectics can in the main be put down to the fact that Engels — following Hegel's mistaken lead — extended the method to apply also to nature.[20] Thus Lukacs was to set up the critical camp in relation to Engels, while it was for Lenin, on the basis of *Anti-Dühring* (rather than the *Dialectics*, which appeared after Lenin's death) to lay the foundations of what was to become Soviet 'Diamat': the petrified mechanical interpretation of dialectical materialism which was to form the philosophical core of Soviet theory, and, formalised as orthodox, official Marxism, was to close down the theoretical prospect for revolutionary practice.*

The relationship of Engels himself to both the critical and the orthodox camp has been obscured. It is not simply a question that, within the *Dialectics* itself, Engels constantly reiterates the fundamental unity of the history of nature and the history of humanity, for that could be a liturgical chant to a god long departed, but that his writing and life work, construed as a whole, testifies to his theoretical and practical recognition of the centrality of human practice in achieving revolutionary change. In its historical context, the Dialectics of Nature is a revolutionising philosophy, a revolutionising way of organising one's understanding of the natural world.

> The whole of nature also is now merged in history and history is only differentiated from natural history as the evolutionary process of *self-conscious* organisms.[21]

> it is precisely the alteration of *nature by men*, not solely nature as such, which is the most essential and immediate basis of human thought, and it is in the measure that man has learned to change nature that his intelligence has increased.[22]

In the *Dialectics* Engels begins by considering the history of modern science, and the part played by the scientific revolutionaries in overthrowing the ideas of a God-ordained and static universe. He sees

---

*This account of the theoretical failure of Diamat for revolutionary practice does not dissolve away the concrete problems facing the young Soviet Union, of famine and encirclement. They too played their part.

science as successively stripping away the ideological layers of a belief in fixity, instead revealing, always more deeply, motion as the central principle of the natural world:

> in the historical evolution of the natural sciences we see how first of all the theory of the simplest change of place, the mechanics of heavenly bodies and terrestial masses was developed; it was followed by the theory of molecular motion, physics, and immediately afterwards, almost alongside of it, and in some places in advance of it, the science of the motion of atoms, chemistry. Only after these different branches of the knowledge of the forms of motion governing non-living nature had attained a high degree of development could the explanation of the processes of motion represented by the life process be successfully tackled.[23]

Thus dialectical analysis stresses two things: first, that all of nature is interconnected into a totality of mutually reacting parts, and hence that no part of nature can be considered in isolation; and second, that each level of description of nature, corresponding to the different sciences, of physics, chemistry and biology, has its own appropriate forms of motion — the more complex or 'higher' forms contain, but cannot be reduced to, the lower. Motion was seen as the inherent attribute of matter, whereby everything in the natural world was in a constant process of transformation. Thus Engels quoted approvingly the Greek perception of the universe of 'all is flux' contrasting this unfavourably with the fixed and constant universe of the schoolmen, where all of nature had its God-ordained place and was in it.

Motion and its laws of the dialectic are not, however, confined to the natural world, but are seen to extend over to the human. Thus where for Marx human practice constantly transforms the natural world, thereby linking its history with human history, for Engels motion — independent of human practice — is to comprehend all change: 'Motion in the most general sense, conceived as the mode of existence, the inherent attribute, of matter, comprehends all changes and processes occurring in the universe, from mere change of place right to thinking.'[24]

Replacing the concept of practice with that of motion is the first step along a path which denies humanity any active part in transforming itself, making it instead merely the puppet of the mechanical laws of nature, and hence history. It was down this path, the path of Diamat, that orthodox Marxism was to travel to the ideology which claims to be *above* ideology, the ideology of 'the scientific and technological revolution!' In it, natural science, seen as the more or less mechanical revelation of nature's laws, becomes neutral and is placed above class; the scientist himself also ceases to have a class position, and instead, as nature's and the Diamat's agent, conducts a revolution within his test tubes. The battlefield of revolutionary class struggle is pushed aside by the revolution of technique.

For Engels, the forms of motion of matter are transformed through the workings of the laws of the dialectic which he formulated on the basis of his and Marx's thoughts: 'the law of the transformation of quantity into quality and *vice versa*; The law of the interpenetration of opposites; the law of the negation of the negation'.[25] In what he himself recognised as an encyclopedic attempt (although he distinguished his dialectical encyclopedia from those of the eighteenth-century Enlightenment), Engels surveyed the whole of late Victorian science in order to demonstrate the workings of the dialectical laws. He sought to draw on the findings of contemporary physics and chemistry to prove the dialectic, while at the same time calling on the dialectic to judge the findings of chemistry and physics. In this way he uses an empirical criterion to validate the dialectic and a dialectical criterion to assess natural science. Thus when speaking of magnetism he claims: 'Dialectics has proved from the results of our experience of nature so far that all polar opposites in general are determined by the mutual action of the two opposite poles on one another.'[26] A few sentences later he dismisses alternative hypotheses on the basis that 'the impermissability of such assumptions follows at once from the dialectical nature of polar opposites'.

It was this confounding of dialectical and empirical reasoning which led to the pursuit of the dialectic in biology, physics, chemistry, and other fields in the 1930s by, for instance, the biologists Haldane and Prenant, a pursuit which could only be sterile. Moreover, while Engels rightly criticised Hegel for forcing the laws of thought on to nature, he himself elsewhere falls again and again into the Hegelian trap:

> (Hegel's) universe, willy nilly, is made out to be arranged in accordance with a system of thought which itself is only the product of a definite stage of the evolution of human thought.[27]

But for Engels

> Nature is the proof of dialectics, and it must be said for modern science that it has furnished this proof with very rich materials

increasing daily, and thus has shown that in the last resort, nature works dialectically and not metaphysically.[28]

Engels thus 'completes' Marx's materialism; the philosophical task is over, and all that remains is to apply the dialectic, for everything in reality is 'already known'. All material existence, both natural and historical, merely has to be fitted in to what becomes a metaphysical category. Thus within the interpenetration of opposites is included: positive and negative electricity, north and south poles; class struggle, identity and difference, cause and effect; and chance and necessity. Sensuous human practice, revolutionary practice, is placed on a par with the movement of molecules. It is as if Engels' recognition that the history of nature began prior to and is likely to end later than that of humanity, pushes ephemeral humanity from the centre of the stage, to be replaced by the metaphysic of nature. Human practice, which for Marx continuously transforms nature, including human nature, becomes passively determined by the motion of matter, a mere 'subjective reflection of objective dialectics'.

## COMMUNISM AND THE RECONCILIATION OF HUMANITY AND NATURE

But to argue for 'forgetting' Engels because of the errors in the *Dialectics* is to refuse to see the overreaching revolutionising philosophy of his writings, and to ignore his own continued concern with the actual theory and practice of making revolution. In the last analysis it is not really a question of whether positive and negative electricity are examples of Hegelian negations, but of how shall we organise our understanding of natural relationships, and whether some understandings serve one kind of general consciousness and other understandings serve another. For both Marx and Engels – despite many of their subsequent interpreters – saw quite clearly the vision which inspires and illuminates the writings in these books, that it is only humanity, and the human practice of communism, which brings the reconciliation of humanity with itself and with nature.

> This communism, as fully and developed naturalism, equals humanism, and as fully developed humanism, equals naturalism; it is the genuine resolution of the conflict between man and nature and between man and man.[29]

# 2

# The Incorporation of Science

## Hilary Rose and Steven Rose

The capitalist mode of production requires continuous innovation in all spheres of life, the creation of new commodities, new technologies, new ideas and new social forms. It is the business of natural science to aid in this process of innovation. Thus under capitalism natural science acts as a direct productive force, continuously invading and transforming all areas of human existence. Marx himself saw that nineteenth-century science acted both as a direct force of capitalist production and also as a means for social control — for the maintenance of the capitalist order. Yet these roles were only partially visible and immanent in nineteenth-century science. It is the thesis of this chapter that, from the mid-twentieth century on, the twin roles of science as a force of production and of social control have become both dominant and manifest, and that this transition is linked with a change in the mode of the production of scientific knowledge, from essentially craft to industrialised production. This change in the mode of production of science has developed over a long period, with some branches of science, such as chemistry, becoming industrialised in the nineteenth-century, and some still to fully undergo the transformation, but from 1945 onwards industrialised science has been the dominant mode.

Thus the major forms of activity included as 'science' are the generation of knowledge and techniques geared to two broad areas of social existence: production and social control. Production science is science for profit, science for the accumulation of capital; conducted in industry, government establishments, polytechnics and universities, it is concerned with developing industrial capacity, exploiting new materials, increasing profitability. Social-control science takes two forms: it is related either to defence against potential external enemies, or the development of techniques for the pacification, manipulation and control of the indigenous population. Even a cursory examination of the annual 'science budgets' of advanced industrial countries such as Britain or the United States makes clear that over the last two decades, between 75 and 90 per cent of the annual total can be embraced under these two heads (77 per cent in Britain in 1974–5, 80 per cent in the United States in the fiscal year 1975). Sometimes, it is true, they are not readily distinguished (for example space and atomic research); sometimes what is social-control science so far as government procurement is concerned is production science for the corporations which contract to do it.

Because of the industrialisation of science and its overwhelming orientation towards accumulation and control, science has become ever more closely and directly enmeshed in the machinery of state and government, so that today there has developed a correspondence between the nature of the state and the institutions and content of science and technology. Scientific and technological policy formation is the expression of this correspondence, but the two most crucial sectors, of production and social control, are characterised by intense secrecy, and hence hidden from view. The global activity of science has thus been masked by the concentration of attention on those areas of academic science which are *not* subsumed within these two categories, areas which, it may still be argued, remain relatively (but only relatively) open, such as molecular biology or, possibly, high-energy physics. Yet attention to these areas, which occupy only a small proportion of the scientific work-force, lays open the danger of mistaking the part for the whole, and also of misunderstanding the nature of the production of scientific knowledge today. For at the present time the dominant mode of production of scientific knowledge has become that of knowledge-as -commodity, as a marketable good with a cash value (see the next chapter).

Although the issues raised at this point speak of the social functions of science, these have been far from the concerns of most bourgeois philosophers, historians and sociologists whose professional task has been to explain the growth and activity of science. Despite an interest in the 1930s in 'externalism', that is the attempt to provide a sociological explanation for the development of science,[1] the dominant paradigm for all three disciplines has been 'internalist'. The scientists were given no social existence: they need no means to produce their non-material or material commodities; all that mattered were the conjectures and refutations between competing individuals and competing schools. Ideas were thus autonomous, unconnected with the social order. Certain kinds of questions became unasked and unaskable. That modern science was born in late Renaissance Europe and not early China (which looked in principle a more likely society, as Joseph Needham points out)[2] requires either a materialist explanation involving the level and kind of economic and social development, or an internalist explanation turning on 'chance', or a racist assumption of the inherent intellectual superiority of Western man.

However, while academic theorists of scientific development were internalist, scientists and 'policy-makers' were hurried along by history (and particularly the Second World War) into a pragmatic externalism. While Popper sought to explain how better theory drove out worse theory, scientists and 'policy-makers' were, in a more Napoleonic vein, trying to secure supplies for their armies of research workers, and to plan the best strategies and tactics of advances.

In order to examine the relations of state and science in the context of the present crisis, it is thus necessary to look in more detail at both the 'theory' and the 'practice' of science policy over the past decades.

## SCIENCE POLICY: PRAGMATIC EXTERNALISM

In its present form, the literature of science policy and the inter-relations of science and government is a mushroom growth of the last two decades, although most of it draws implicitly on Bernal's *Social Functions of Science* published in 1939.[3]

Bernal's work pioneered a Marxist analysis of twentieth-century science. He recognised the necessity of scientific innovation for capitalism, and the role that industry and the state played in funding and directing the advance of science. Science thus generated profit, and was used in the maintenance of state power. None the less, despite the fact that science was integral to capitalism, ultimately it was in contradiction with it; capitalism continuously frustrated the potential of science for human good — for the elimination of disease, poverty and toil. Immanent within capitalist science — with all its distortions and its ideological role — was a good science, seen as harmonious with socialism. The contradiction between science and capitalism showed itself in the incapacity of capitalism to invest adequately and plan for science. Socialism above all meant planning, the rational development of science for the benefit of the people. He looked to 'the new socialist world' — to a 'science for the people'.[4] Yet a one-sided reading of Bernal could claim that it was only necessary to plan, to increase the scale of scientific spending some tenfold (to the level of that current in the Soviet Union for example) and all would be well. The ideology of techno-economism began to take shape. In a sense the British Labour Party's election manifesto of 1964, concerning the forging of socialism in the white heat of the technological revolution, was a debased after-echo of the Bernalian analysis of twenty-five years previously.[5]

The expansion of science-policy studies, especially in the United States, was to postdate Bernal's book by fifteen years and Hiroshima and Nagasaki by a decade. The reasons for the concern of the state with science in the late 1950s and 1960s are clear; they are symbolised by the growth of the science budget along its exponential path to the magic 3 per cent of G.N.P. in the United States, the Soviet Union and Europe. As politicians attempted to understand, control and direct this growth, political scientists, from D. K. Price[6] through Gilpin,[7] Barber[8] and Schooler,[9] wrestled with questions of the relationship between knowledge and power. Whilst some of these accounts were merely hagiographic, critics such as Lapp[10] and Skolnikoff[11] responded to the issue of human survival posed by nuclear weapons by urging the democratic control of science. But such liberal responses, lacking a clear theoretical understanding of the relationship of science to capitalist production, could generate no more than mythical solutions, framed inside the theory of countervailing powers within the bourgeois state. For the politicians, the argument took the implicit (and sometimes explicit) view that it was possible and desirable through the funding of science to influence the direction and speed of scientific advance. While rarely was this pragmatic externalism so crude as to suggest that the frontiers of knowledge could be extended by the mechanical application of resources to a particular problem, they were none the less seeking in practice to promote general lines or even to achieve defined goals; the most conspicuous examples of this have been the Kennedy response to Sputnik — an American on the moon within ten years — and now the cancer and heart programmes, or, on a lesser scale, the sickle cell anaemia programme. While during the peak of the high-spending years both technological and pure-science projects were dreamed up and sponsored with reckless abandon (Greenberg[12] was able to document some of the more notorious projects in pure science in the United States such as the Mohole or the Linear Accelerator), as the economic climate chilled a greater financial and scientific caution prevailed. None the less particular areas of 'pure' science have been seen both by governments and industry as integral to their political or economic purpose. Thus governments have given high-energy physics massive state support for both national and international facilities.

While at the ideological level 'pure science' is divorced from 'technology', in practice, whether that of the state or industry, the two are intimately connected. Industry invests in 'pure science' as well as technology; as witness Bell Telephone Laboratories' funding of William Shockley's work which both won him a Nobel prize and laid the foundations for the commercial exploitation of the transistor and microelectronics. As J. J. Salomon, of the Organisation for Economic Co-operation and Development, put it, 'the field of pure research constitutes the extreme case of coincidence between the interest of power and the interests of knowledge'.[13]

Thus, during the 1960s, governments sponsored a series of inter-disciplinary research groups within universities, at national and supranational (for example at OECD, and UNESCO) levels. These worked on 'science studies', 'science of science', 'science policy', 'science and government', according to the ideological perspectives within which the questions were to be posed. The emphasis (within the framework of techno-economism) was on quantification, and, as the increasing costs of science and the inevitable slackening off from the exponential growth phase occurred, attempts to justify expenditure on science by proving its relationship to economic growth. This has been a constant feature, for instance, of British work, from Carter and Williams (advisers to the British Labour Government)[14] to Freeman's OECD-sponsored work at Sussex.[15] In the United States, the most conspicuous example was the exercise mounted by the Department of Defense (DoD) under the name 'Project Hindsight'. Most of the research produced by these groups has been empirical, and, at the same time, separated from the thrust of the necessity felt by politicians or industry, it has yielded for the most part neither theoretical explanation nor even facts very helpful to the state's policy-making needs. In the United States, which was the first to sponsor such research groups, there is evidence of the withdrawal of funding, which suggests that so far as their sponsors were concerned, they failed to deliver the goods.

It remained the case that when members of the scientific elite discussed science policy, their peers accorded them an interest denied the research units. Thus the book *Criteria for Scientific Choice*[16] written by Weinberg, physicist and head of Oak Ridge National Labora-

tory in the United States, received a double welcome; for its suggesting
of rules of procedure for establishing research investment priorities, and
pragmatically balancing internalist and externalist criteria. Even useless
contributions from this stratum were listened to with courteous respect.

None the less, by the 1970s, it has become apparent that the achievements of science policy have been limited. This was recognised at the
political level, as many governments abandoned their 'Minister of Science' or 'Science Adviser'. Science policy-making has reached something
of an impasse, the heady days of scientific expansion itself are now well
and truly over. The 1960s and early 1970s saw, as we will discuss in
more detail in the case of Britain below, the deepening integration of
science into the state. Where scientific rationality pervades all aspects of
social life, there is no longer any need to set science apart and demand
for it a policy as if it had an autonomy, as if biology and physics were
to be regarded as something separate from an extension of medicine
and agriculture on the one hand and military and industrial development on the other. In this same vein, all those many books entitled
'Science and Society' (also including one by authors who should perhaps
have known better) are, as Levy Leblond and Jaubert observe,[17] sociologically incorrect in that they juxtapose, as distinct even if interacting,
scientific and social systems.[18]

## THE UNITY OF SCIENCE AND TECHNOLOGY

The point is that, despite the paradigm within which bourgeois historians, philosophers and sociologists of science have operated, modern
science and technology are indivisible. The particular character of
modern science ushered in with the Galilean revolution is precisely that
it is directed towards experiment, use, technology itself; it is this
which sets modern science apart from that of classical Greece, Babylon
or India. The contemporary production of scientific knowledge is
predominantly through the method of experiment, inherently committed to acting on the natural world, in order to understand and control
it. At the level of consciousness of individual scientists, a quite contrary
view was commonly expressed from the nineteenth through to the mid-
twentieth century. This emphasised the disinterested and non-utilitarian
nature of the work of the 'man of science'. Often quoted are the
examples of the mathematician G. H. Hardy, who claimed that what he
liked about mathematics was that it was of no use to anyone, and of
Ernest Rutherford, who, laying the foundations of nuclear physics at

the Cavendish, was to say that he saw no application for his work. Their belief that they were pursuing knowledge for knowledge's sake savours more of the social functions of pre-modern science, where science is on a par with other intellectual and aesthetic activities such as music or poetry, than those of contemporary science. (The argument here is not that there is *no* science without a social function; it is that the dominant mode of production of scientific knowledge has social functions.) In the case of Rutherford, the relevance of his pure science to military technology was to become all too clear despite his disclaimers. The significance of the Hardy–Weinberg equation for population genetics weakened, if less dramatically, Hardy's claim of uselessness. What began in the nineteenth century as a technical division of labour between the development of theory and its application to particular practical problems became increasingly confused with a social division of labour. The distinction between 'pure' science (and the word 'pure' is not without its sociological connotations) and 'applied' (impure, 'dirty' science) was and is sustained by the social institutions of science. The elite status of the non-manual scientists as against the manual engineers has lingered on in a way which would satisfy Pythagoras himself (see the chapters by Gorz and Cooley).

It is perhaps worth reminding ourselves that this social division of labour was neither considered necessary nor practised by the founders of the Royal Society. As R. K. Merton points out,[19] their researches ranged freely both over those questions which had a primarily theoretical interest and those which were primarily practical. Robert Boyle was, for example, well aware of the connection between his discovery in 1662 that the volume of any gas varies inversely as the pressure at constant temperature, and problems of interior ballistics. For that matter, long before, Leonardo da Vinci had sought aristocratic patronage for his scientific research on the grounds that this would yield improved weaponry. So did Archimedes. However, the economic base does not determine the superstructure in any mechanical way, so that once theoretical questions are posed, the development of theory takes on a life of its own. In the context of Britain's industrial revolution, theory and practice, in terms of scientific theory and industrial innovation, became for the most part institutionally separated.


### THE STUDY OF SCIENCE – ACADEMIC INTERNALISM

How has bourgeois history, philosophy and sociology of science come to ignore the unity of science and technology? We can see this in the case of a leading sociologist of science, R. K. Merton, whose early work, *Science, Technology and Society in 17th century England* is a rejoinder to Hessen, a Soviet physicist who, as part of the Bukharin-led delegation to the International Congress of the History of Science and Technology held in London in 1931, presented a classical Marxist thesis of scientific growth.[20] Hessen took Newtonian mechanics and showed how it was developed directly in response to the needs of burgeoning capitalism. Whilst his internalist British critics at the meeting sought to correct Hessen on small points of 'fact', Merton responded to the theoretical challenge of what was to be called the 'externalist' theory of scientific growth. Merton, in exploring the influence of Puritanism in the development of seventeenth-century science, stands to Hessen as Weber's Protestant Ethic can be juxtaposed to Marx's theory of the growth of capitalism. Merton attempted to show that science develops not solely in response to economic needs, but also requires a supportive value system — namely Protestantism. While this comes close to arguing that the superstructure — in the form of religious ideology — determines the base, Merton was concerned to examine the base/superstructure relationship. However, the emphasis on religious ideology and its compatibility with the scientific ethos pushed the work away from any economic explanation into a form of sociological internalism, characterised by a preoccupation with science as a more or less autonomous subsystem. This preoccupation with the scientific ethos was paralleled by the philosopher Polanyi's conception of the scientific community as a self-governing collectivity.[21] This variant of internalism, which dominated the academic sociology of science for thirty years, ceased to address itself to questions of the interpenetration of science and the social order at the cognitive level, or even of scientists and the social order at the structural level. Instead, it interviewed Nobel Laureates, phage biologists, topologists, high-energy physicists, and so forth, and assumed that the productive activity of the whole of science, including the industrial and military, could be interpreted through the study of the elite.

Thus the fundamental character of science and technology in their social functions was lost to sight.

At the very moment when the advent of war unequivocally destroyed the foundations of Polanyi's thesis, which was an effort to preserve science from the dreaded Marxist planners headed by J. D. Bernal, sociology was to conceive this curiously backward-looking conception of science. Polanyi's — and hence Merton's — community of self-governing scientists was destroyed in a real world which no longer asked *should* science be planned, but *how* should science be planned; not *should* science serve the state, but *how* should science serve the state? During the Second World War, even erstwhile *laissez-faire*-minded scientists energetically joined the Bernalists and the Zuckermen to plan science, mobilising scientists to work on radar, bombing patterns, CBW and of course, atomic weapons. The message was stamped home in the Manhattan project, that uneasy alliance of J. Robert Oppenheimer and Major-General Groves, which was the embodiment of the most massive intervention of the state into science (and, for that matter, of scientists *vis-à-vis* the state) and yet took place within that most arcane of sciences, physics itself.

## THE MANAGEMENT OF BRITISH SCIENCE

In Britain the transition between *laissez-faire* and state-managed science is symbolised by the contrast between two seminal government reports on the relationship between state and science, separated by more than fifty years, years of which the Polanyi – Bernal debate and the Manhattan project mark the midpoint. The two were those of Lord Haldane in 1918 and Lord Rothschild in 1972.

The *Machinery of Government* Report[22] was the culmination of a series of studies embarked upon by Haldane and his colleagues in the war years. It established the Research Council structure by which 'pure' research has been funded in Britain ever since. The point for Haldane was very clear. A modern state needed science in order to survive — just how much had been shown by the failures and limits of British technology against a Germany superior in chemistry, physics and engineering in the early war years. Haldane did not need the elaborate exercises of later policy-makers, such as the DoD's *Project Hindsight*, to be confident of the links between basic science and defence and industry. Yet it was also apparent to him that these links were much mediated; to be effective, science and scientists needed space and protection from immediate production and military pressures. So the famous Haldane principle was enunciated, by which the Councils were independent of the Departments of State which might be expected to be affected by their research (although such Departments were expected to conduct their own research as well). The advantage of this independence, as Haldane put it, was as follows:

It placed responsibility to Parliament (for the research) in the hands of a Minister who is in normal times free from any serious pressure of administrative duties, and is immune from any suspicion of being biased by administrative considerations against the application of the results of research.[23]

Thus the 'space' available for scientific autonomy was carefully defined and protected in the interests of efficiency, and the very space itself became hallowed as the cornerstone of scientific freedom until the end of the Second World War. From 1945 onwards, successive governments (whether Labour or Conservative) drew the net of state – science interaction tighter, culminating when, under the 1970 Conservative government, the Rothschild Report, *A Framework for Government Research and Development*[24] challenged the Haldane principle head on, and, over the vociferous protests of the scientific elite, was accepted as the future basis for the management of science.[25]

What Rothschild did was to state clearly that research and development were not autonomous but had a purpose to be defined by the state and its industrial counterparts; policy decisions were not the free prerogative of elite scientists merely because they were scientists – except in so far as they also happened to hold other governmental or industrial positions (see below). Even the language of the Report was in a style which made the new relationship manifest; Hardy and Rutherford might turn in their graves.

> This report is based on the principle that applied r and d, that is, r and d with a practical application as its objective, must be done on a customer – contractor basis. The customer says what he wants, the contractor does it (if he can): and the customer pays.

Rothschild hereby makes explicit the extent to which managed science has, under modern capitalism, become part of a managed society, but a society whose management is itself scientifically founded. It is from this extension of scientific rationality, characteristic of the present phase of capitalism and state socialism, that the conception of science as first the domination of nature, and then the domination of humanity, has been born.


THE DOMINATION OF NATURE AND THE DOMINATION OF HUMANITY
The elision between knowledge and power which had existed on a relatively *ad hoc* basis – close in time of war, neglected in peace – had

thus by the mid-twentieth century become writ large and institutionalised. However, what had happened in this apparent fulfillment of the Baconian vision was that human progress, which was the underlying goal of New Atlantis, became replaced by technical progress. Thus, writing in the defence of Operation Plowshare; a proposal to use nuclear explosions to blast out deep harbours, the father of the American H-bomb, Edward Teller, described science as progress, and progress 'cannot and will not be stopped'.[26]

Where in the past the ideology of science proclaimed its socially liberatory function (much as official Soviet science still does — see below), Nagasaki and Hiroshima marked unequivocally what had only been glimpsed before, the alliance between the domination of nature by science and the domination of humanity by power.

Partly because the Bomb had been so devastating, the scientists, as an elite, were unable to sustain the ideology that science and technology were socially progressive, yet assumed that they carried particular political responsibilities in that they believed that the research they did, unlike that of the historian or artist, reacted very directly upon society. At the same time, there was a danger that the Bomb would be seen as an *inevitable* result of physics, so that anyone with qualms concerning its use would have to stop doing physics; the responsibility would have been too much (indeed many did go into biology for these reasons). The defence against this criticism was to claim the *neutrality* of science, as a force either for good or evil depending upon the whims of society. It was merely the *application* of science which was non-neutral. This convenient conjuncture enabled many scientists in the two decades that followed to continue simultaneously to do high science — even accepting research grants from the military to support it — while at the same time professing radical political attitudes or arguing against particular developments in the arms race; the discord between the objective reality of the uses of science and the consciousness of the scientists became almost complete.

The critique of science as inherently geared to the domination of nature and thence of humanity had been developed by the neo-Marxist Frankfurt School of critical social theory. Unfettered by any allegiance to the particular form of socialism emerging within the Soviet Union and subsequently within Eastern Europe, members of the Frankfurt School were able to continue the critical examination of the nature of alienation under increasingly corporate capitalism and bureaucratic socialism. Retaining their interest in the psychological dimensions of alienation, an interest elsewhere dismissed by orthodox Marxism as the concern of the early Marx, the School, and most notably Hork-heimer,[27] Adorno,[28] Habermas[29] and Marcuse,[30] was able to explore the ways in which the impersonal rationality of science extended into the political process itself.

The implication of this analysis is that, as it has developed under the conditions of industrialisation in Western society, science has been concerned, not so much with ensuring the vision shared by Marx and Engels, of men and women living harmoniously with nature, but, instead, first with the control and manipulation of nature, and then of one another.[31] This can be seen at many levels; from the choice of words used to express their activities by technological ideologues (the *conquest* of space, the *modification* of weather, the *exploitation* of natural resources, the *control* of brains and behaviour); to the recognition of all the unintended consequences of such technological advances, from the ecological hazards of dams to the drug-based medical practice of Britain or the United States. Where scientific and revolutionary optimists like Bernal had shared an inevitably progressive view of science (beautifully captured by Richard Gregory, the editor of *Nature* in the 1930s, who said 'My grandfather preached the gospel of Christ, my father preached the gospel of socialism, I preach the gospel of science'), for the more pessimistic Frankfurt School it was not merely the *relations* of technological production which were oppressive but the very nature of the technological product itself. Science and technology under these conditions could not *but* be oppressive. What is more, this oppression is not merely reflected in their products, the hardware of social-control technology, from napalm and fragmentation weapons to ritalin and psychosurgery, but also in the ideological role played by scientific theories in the preservation of the existing social order, from Social Darwinism in the nineteenth century to IQ testing or ethological theories of 'innate' forms of human society today (see Chapter 7).

This is clearly seen in Marcuse's *One Dimensional Man*, which portrays a social world pervaded by technological rationality. Marx had argued that the antagonism between nature and man, and hence man's alienation from nature, including his own nature, was integral to capitalist society. Natural science, at least for Engels in the *Dialectics of Nature*, contained an enlightenment concept of scientific progress as equalling human progress. Marcuse, in a century of the massive growth in the scale and power of science, discerns science and technology as a particular mode of rationality aiding human oppression, either directly as the technology of repression, or individually through biological manipulation: 'Technology seems to institute new, more effective and more pleasant forms of social control and social cohesion.'[32] Thus political questions are dissolved into technical issues to be resolved by experts. Technological rationality becomes political rationality. Critical opposition is denied legitimacy. While the Frankfurt School recognised the immense power of technological rationality, their isolation from political practice led them to pessimism and a sense of impotence. They could not conceive of the theoretical and practical possibility that even the most monstrous technology could prove to be a paper tiger. For them, the victory of the Indochinese people could not occur.*

### PLANNING SCIENCE FOR HUMAN LIBERATION

The orthodox Marxist optimism about the inevitability of the relationship between scientific progress and human progress fails when confronted with the reality of the incorporation of science in capitalist societies over the past thirty years. Yet the Indochinese victory makes clear that the opposition to this optimism, the Frankfurt School's pessimism over the inevitability and incontestibility of technological oppression, is equally in error. We can see these contradictions working themselves out in both the Soviet Union and China, following 1917 and 1948.

In the Soviet Union, the linking of science to social needs and state interests has been explicit and theory-based since 1917. Science, belonging to both the economic base and the superstructure, was to play a leading part in the achievement of human liberation, first through socialism, then communism. In the early 1920s, during the period of the New Economic Policy, it was considered enough to facilitate the work of the engineers and fund the scientists on an increasing scale without too close a control of either their ideology or products. Scientists were divided between bourgeois researchers trained in pre-revolutionary Russia, and young Marxists who had been trained by them but who were working in — and for — the post-revolutionary society. The debate was conducted by reasoned argumentation — by critique and counter-critique. This style of conflict prevailed until Stalin imposed his own intellectual and cultural solution to this and

---

*This is not completely fair to Marcuse, who, unlike the other members of the school, responded optimistically when the various movements for liberation surged forward in the late 1960s.

many other areas of theoretical dispute.

The questions opened in many areas of science, of which the most notorious was genetics, were not only concerned with the battle of ideas, between for instance, the 'bourgeois' paradigm of genetics, with its concept of fixed inheritance and chance mutation, and the 'socialist' paradigm of infinite plasticity and environmental modification. They also concerned the class origins of the scientists themselves. The deep-bedded socialist view that knowledge is forged in practice favoured Lysenko, the peasant and practical plant breeder — in this sense a cultural revolutionary — over the aristocratic geneticist Vavilov (see *The Radicalisation of Science*, Chapter 2).

Leaving to one side the actual historical outcome, what we can see in the early phases of these struggles was the spontaneous attempt at a cultural revolution, but an attempt *without* the clear understanding of the role of mass participation characteristic of Maoist theory. The attempt, therefore, became the victim of Stalin's administrative politics whereby both redness and expertise were to be defined from the top. Thus, although from the Great Break of 1927 onward[33] there was an attempt to link both the goals of technology and the ideology of the technologists themselves with the needs of communism, beginning with the engineers and spreading by the 1930s and 1940s to all branches of science,[34] both the redness and expertise were themselves trapped in an increasingly bureaucratic and hierarchical society. In any event, the attempts to look for specifically proletarian forms of science have been modified in the last two decades into an assumption of the automatic elision between the needs of the Soviet state and the advances of a science seen as neutral or beneficient.

It has been left to the Chinese revolution to advance the theory and practice of cultural revolution. Starting in 1951 from the Leninist perspective that 'a scientist or engineer would come to accept communism through the data of his science and in his own way', Maoist theory had added the need for ideological remoulding of intellectuals.[35] The Communist Party was faced with the double task of developing Chinese science and at the same time harnessing it to the needs of the people, particularly in production. The emphasis on production to be achieved through bureaucratic professionalism (Liu Shao Chi-ism) led to the perpetuation of hierarchy and the dominance of theory over practice. The cultural revolution, beginning in 1966, returned to the question of ideological remoulding, but in a much more root-and-branch way, challenging bureaucracy and restoring the classical practice – theory –

practice epistemology of Marxism, so that all forms of leadership which
had sought to derive their authority solely from expertise were heavily
criticised. Horn[36] and others have discussed the changes in the
biomedical sciences, reversing the drift to urban and hospital-based
sophisticated technology beloved by medical technocrats, towards a
rurally orientated system into which traditional medical methods otherwise 'invalidated' by high technology have been integrated. Here,
authority is based on service to the people expressed through expertise.[37] We know less from first-hand accounts of the workings out and
successes and failures of the scientific laboratories in the attempt to be
both red and expert. Although there have been continuous attempts to
proletarianise the laboratories, to make academic scientists learn from
workers' experience, issues within science itself appear, on the whole,
to have been dealt with by experiment and debate on the 'hundred
flowers' principle rather than by repeating the disaster of the Lysenko
experience. It is thus possible for philosophers of science to be either
'Lysenkoists' or 'Mendel–Morganists',[38] although the failure of the
Lysenkoists to get their theories to work out in practice is recognised.[39] Stalin's solution to the problem of genetics was, from a very
early period, considered by Mao and the Party to be an incorrect way
of handling scientific questions:

> Letting a hundred flowers blossom and a hundred schools of thought
> contend is the policy for promoting the progress of the arts and the
> sciences and a flourishing socialist culture in our land. Different
> forms and styles in art should develop freely and different schools of
> science contend freely. We think that it is harmful to the growth of
> art and science if administrative measures are used to impose one
> particular style of art or school of thought and to bar another.
> Questions of right and wrong in the arts and sciences should be
> settled through free discussion in artistic and scientific circles and
> through practical work in these fields. They should not be settled in
> a summary fashion.[40]

However, Mao makes clear that this is no simple adoption of cultural
liberalism; the task remains, within the policy of one hundred flowers
in science, to ensure that they are socialist blossoms and not capitalist
weeds. The chief criteria by which science and art are to be judged are
political, emphasising the need for culture to unite and not divide the
people, to strengthen the leadership of the party, to help socialist
transformation and to contribute to international socialist unity. In

addition, but subordinate to the political criteria, there must be appropriate technical criteria. But as Mao concludes: 'In a socialist country like ours can there possibly be any useful scientific or artistic activity which runs counter to these political criteria?[41]

By contrast, in orthodox Soviet writings today, such as the essays by Mikhail Millionschikov (vice-president of the U.S.S.R. Academy of Sciences) and others in *The Scientific and Technological Revolution*[42] science and technology are seen as value-free (that is, there is no specifically socialist biology, and the explosive debates of the Lysenko period are dismissed as misguided). However, the maximisation of the potential of science and technology for human welfare can only take place within the framework of the Soviet system. It is taken for granted, therefore, that under Soviet socialism state and science are in intimate correspondence by way of an elaborate policy-making machinery, and, indeed, that bureaucratic professionalism is an efficient and desirable way to perpetuate this correspondence (see, for example, Kapitsa on the question of how the scientific elite should be recruited).[43] However, in order to maintain *both* that science is neutral and that it corresponds with the needs of the people in a beneficent Soviet society, it is necessary to argue that in capitalist societies, state and science are in contradiction. As Millionschikov puts it, 'the principle of private enterprise in the age of nuclear energy, electronics and cybernetics will become increasingly exposed as historically obsolete'. Scientific rationality — without being made over to a specifically socialist science — is thus supposed to expose the irrationality of capitalism. It is not necessary to be Marcuse, or for that matter to work in Solzhenitzyn's *First Circle*, to see the falsity of such a proposition.

THE INDUSTRIALISATION OF SCIENCE

The economic base of the Soviet state socialist system and that of Western capitalism are quite different; none the less, in both societies there has been an incorporation of science into close correspondence with the technological and ideological needs of state and industry. In this sense, the critique of science as the domination of nature and of its oppressive role may also be generalised as describing the situation in both the Soviet Union and Western capitalist countries.

In addition, the changed mode of production of scientific knowledge has resulted in a shift in the internal organisation and social relations of science. Seen most graphically in Big Science where a whole laboratory may sign one short paper, this change has been widely recognised as the shift from craft production to the industrial production of knowledge.[44]

Seen most graphically in Big Science where a whole laboratory may sign one short paper, this change has been widely recognised as the shift from craft production to the industrial production of knowledge.[44] Where scientists once worked as individual producers of knowledge, now they work in large, hierarchically organised teams characterised by an increasingly intense division of labour. Each scientist, or more accurately each scientific worker, for so the rank and file of the scientific factories of today must be termed, has fragmented partial skills, bound to a purpose only fully understood by the project director and those who set the goals of the group of the laboratory. The scientific worker, as Gorz and Cooley's chapters make clear, has become a mere detail labourer.

The process has continued and expanded, science by science, beginning with chemistry, through physics to molecular biology. Scientific workers are like factory workers, defined by the machine they tend, so that we have lathe operators and nuclear magnetic resonance (n.m.r.) operators, typists and computer programmers, n.c. machine operators and spectroscopists. In the nineteenth century Whewell gave recognition to the gentlemanly nature of the pursuit of knowledge with his phrase 'cultivators of science'. Twentieth-century professionalism was marked by the term 'scientist'. Now with the industrialisation of science, a proletariat has emerged, the rank-and-file *scientific workers*. Scientific workers in industrial and governmental research establishments have been found to be indifferent to the norms of science and instead are preoccupied by conditions of work, pay, security and prospects.[45] The alleged commitment of 'scientists' to 'Public Knowledge',[46] where the reward system gives acclaim through peers, eponymity or prizes, thus remains irrelevant in most of the research system. Despite — or perhaps because of — this transition from scientific community to scientific factory, there remain, leading, planning and administering science, a scientific elite who are happy to share and perpetuate the myth of the self-governing community of scientists. It is only these, as we have said, who have been 'visible' so far as much of the academic sociology, philosophy and history of science are concerned. It is only these whom Shils, contrasting scientists with the 'laity', urges to keep faith[47] and who share in Monod's conception of science as humanity's noblest activity.[48] The restricted membership of this elite stratum is evident in the management of the entire research system. In the United States, for example, it has been estimated that some 200 - 300 key decision-makers — primarily scientists — constitute the inner elite out of a total scientific work-force of some two million. For the rank and file of scientific workers, alienation is the norm, but with the double burden imposed by an ideology which insists that within the pocket of every laboratory worker's coat lies the gold medal of the Nobel Laureate.

## DOMINATION OF NATURE OR PAPER TIGER?

The disenchanted view of the interpenetration of science and society is increasingly widely held. It is not only the politicised Marxists or anarchists who have felt and attempted to analyse the oppressive character of science and technology in contemporary society. Together the domination of nature and bureaucratic rationality are seen by many to represent an almost irreversible, increasing and potentially disastrous trend. The domination of nature threatens to become the total destruction of nature. Yet despite the apparent similarity of the diagnoses, the prescriptions are very different. Some conservative theorists, like Ellul,[49] make almost no prescription at all — to understand is sufficient. Others — for example the Club of Rome — propose that the self-same social strata which created the problem in their own interests should now be relied upon to solve it — in their own interests. The counter-culture anarchists, such as Roszak,[50] seek to return to some golden pre-industrial age, which for the masses can only mean privation and premature death.

The social function of these 'solutions' to the crisis is to increase political apathy and confusion. Ellul adopts a stoic pessimism which represents the advocacy of fatalistic resignation, the self-interest of the Club of Rome can produce no more than a cynical smile, while Roszak's proposals, impractically absurd for the masses, merely create confusion amongst radical intellectuals. Thus all these 'solutions' have an ideological, mystifying role, and arise from a failure to locate current developments of science and technology within a historical context. Only a clear recognition of the present phase of capitalism and imperialism will make it possible to develop the forms of potential activity which will enable the masses of the people to transcend ideological mystification, recognise the true role of science and technology in *this* society and reveal them as paper tigers, capable of being combatted and defeated, so as to make way for a genuine science for the people.

# 3

# The Production of Science in Advanced Capitalist Society

*Giovanni Ciccotti, Marcello Cini and Michelangelo de Maria*

## THE RELATION OF SCIENCE AND THE SOCIAL ORDER

The dominant faith in the automatically progressive character of scientific work has diminished in many sectors of society, and even partly within the scientific community, since the 1960s, giving way to a growing scepticism concerning the liberatory power of science and the very worth of scientific knowledge itself. This situation is related to the pessimism over the possibility of redirecting towards human ends the enormous growth in productive forces that has emerged with the so-called scientific and technological revolution, and to the identification — empirically justifiable — of the roots of this development in the increasing integration of science and technology into the productive structure of mature capitalism.

Even if scientific and technological research continues, perhaps with some internal readjustments, to serve as the propulsive element of growth, it becomes more and more apparent that it generates a type of growth that is advantageous for a few while being increasingly costly for the many. As a consequence, technocratic and rationalising optimism has given way on the one hand to a storm of anti-scientific pessimism, and on the other to a reaction emphasising a clear distinction between natural and social science. Both these positions are inadequate, and cannot lead to an understanding of the real issue. There are dangerous confusions implicit in attributing to science the responsibility for the anti-human forms presented by modern technological society. However, without denying that the protest against science, instead of searching for the social roots of the problem, often simply takes up themes and suggestions that are derived from decadent late romantic irrationalism, it is not enough to emphasise this danger. Simply to reaffirm the validity of science is to limit oneself to seeking to exorcise the misty ghost of a crisis rather than confronting it on real ground. In other words, it is not enough to reject the contention that dehumanisation flows from the process of scientific – technical reification, without investigating to what extent this reification is reflected in the production of science itself.

The problem therefore is real and cannot be ignored; it is to investigate the links that exist between science as a particular form of social human activity, and the social relations of production that in general regulate the activity of human work in *this* society. This is the significance of moving from a general recognition, now largely accepted, of the 'non-neutrality' of science, to a more precise identification of the various levels and mechanisms of reciprocal interaction of these activities, and of the possible interventions available to transform the social role of science though the explicit recognition of alternative social goals. It is, therefore, only by understanding the profound crisis that throws into doubt the significance, objectives and values of science, that we can overcome the impasse between the anti-scientific pessimism of irrationalism and the naive optimism of abstract rationalism; and it is because of our consciousness of this crisis that it is important to attempt to rediscover in the conception of nature and the scientific methodology of Marx, the instruments to analyse and reconstruct the natural, historical and ideological totality which is our society. It is clear from this terse account that it is our conviction that only within the framework of a correct, even if schematic, reconstruction of the links that bind science to the other structural and super-structural components of this totality will it be possible to provide concrete and not subjective replies to the questions that the crisis raises.


THE DIALECTICAL ANALYSIS OF SCIENCE

There are two aspects of Marxist thought which are prerequisites for this analysis. The first consists of the refutation of the separation of social relationships between humans on the one side, and relationships between humans and nature on the other, into two distinct, rigid and non-communicating spheres. In fact to accept such a premise prevents any possibility of formulating the real problem. This is true also for those who, denying this separation in the abstract, none the less assume nature and history to be two separate 'fields of application' of dialectical materialism, thereby risking falling into a more or less modified version of Stalinist 'Diamat' — that is, an 'ontologicalisation' of the dialectic as 'a conception of the universe, a positive principle of the world',[1] the reverse of what Marx meant.

The correct use of Marxist dialectical thought enables us to avoid a second danger inherent in another widespread interpretation of the relationship between nature and history. This is the tendency to interpret the reciprocal interaction between these two spheres as an unilinear correspondence between 'the growth of the productive forces' (understood as an autonomous process of the growth of the domination of humanity over nature) and of social relationships which tend to adapt themselves to the level of this growth; this falls into a mechanical conception of history and society which cannot be attributed to Marx, in spite of its acceptance within the Marxist tradition.

The second aspect of the Marxist conception of dialectics and materialism to which we wish to draw attention is that which underlines that it is not possible to separate knowledge from practice without reducing the first to mere passive reflection on a given object, and the second to the active manifestation of subjective thought. To maintain a dialectical unity of perception and action, implies rejecting a separation between judgements of fact (passive reflections of an object) and judgements of value (subjective practical activity): a rejection of the separation between science and ideology. This point is essential, we believe, in order to recognise the ideological form in which any kind of knowledge — including what is generally regarded as scientific knowledge — is expressed; namely in order to grasp the implicit plan of practical activity which, in a more or less mystificatory way, is always present in knowledge. In this sense the brief description of the Marxist conception which we have sketched out itself shows why we refer to it as an exemplary case of scientific knowledge. As such it allows one, at least in principle, to reconstruct the unity between ideology and structure, thus demystifying the apparent autonomy of consciousness from the process of material production. In this way it becomes possible to give an account of the significance and the implicit scope of modern science. One of the objectives of the analysis in this chapter is to begin to move in this direction.

Furthermore, it is essential, in our opinion, to make explicit reference to what Marx claims to be the 'correct scientific method' for the 'reproduction of the concrete in the field of thought'. This is the method of political economy, the logico-historical method set out by Marx in 1857 in his introduction to the *Critique of Political Economy*.

On this point, too, there exist within the ambit of Marxism contrasting interpretations which reflect both different theoretical positions and evaluations of the principal contradictions in contemporary capitalist society. Thus Marx claims, in the text we have cited, that:

> The bourgeois society is the most highly developed and most highly differentiated historical organisation of production. The categories which serve as the expression of its conditions and the comprehension of its own organisation enable it at the same time to gain an insight into the conditions of production which had prevailed under all the past forms of society, on the ruins and constituent elements of which it has arisen and of which it still drags along some unsurmounted remnants, whilst what had formerly been mere intimation has now developed into complete significance. The anatomy of the human being is the key to the anatomy of the ape.[2]

It is therefore a consequence of the 'correct scientific method' that we reject a view of the present as merely the point of arrival of a chronological succession of successive levels which have prepared its coming. It is on an analysis of the 'most complex and developed historical organisation of production' that we must base ourselves if we wish to confront the problems of the worth of science and of the social function of research. To assume as the object of analysis instead, science in the abstract, as a generic human activity in which people have always engaged, independently of any particular historically determined form of economic and social organisation, signifies 'considering the real as the result of self-coordinating, self-absorbed and spontaneously operating thought'.[3]

It has been neglect of this fundamental aspect of analysis of reality which has condemned to sterility the development[4] of Engels' programme for science into a solution which finally solves the 'gnosological problem', a permanently valid reference point guaranteed to clear the field once for all from the contingent and changing vicissitudes of the social relationships between humans and to replace it with a steadily deepening and more correct relationship between humanity and nature. We are not concerned to discuss if, or in what measure, Engels was a 'good Marxist' in his analysis of the dialectics of nature. What we reject is the attempt to elevate Engels' analysis, which had

relevance at a particular historical moment, into an unchangeable paradigm which claims to make possible a decision on the merits of any
particular scientific debate.

Science was at that time only partially integrated into the productive
process, but was so strongly conditioned by philosophical thought,
dominant ideas and traditional culture, that national 'schools' corresponding to different levels of social organisation could be identified. It
was through a recognition of the ideology which permeated diverse
scientific theories, and from a materialist and a dialectical point of view,
that Engels was able to analyse the positions taken by Darwin, the
atomists, the organicists, and so on. But Engels himself took account of
the transitory nature of this particular historical situation when he
stated that 'the progress of theoretical natural science will render my
work largely, or completely superfluous'.[5] In attempting to deal with
the same general problem, it is therefore not necessary to try to revive
the same debate, or to reproduce the same interpretive schemes within
today's science, which, as a concrete social human activity, is qualitatively different from that of the previous century. If the differences in
its social goals, the mode in which it is produced, and the ideology
which permeates it, are not taken into account, there is the danger of
relapsing into scientism. Thus on one hand it is theorised that 'it is
precisely in the possibility for man to amplify his own horizons, to
extend his consciousness and domination over those parts of nature
which are not yet controlled, rather than the theoretical viewpoint or
the consequences of this practice, which are at the roots of the progress
of humanity'.[6] On the other, hastily dismissing the condemnation that
Stalin launched against Mendelian genetics and quantum mechanics in
the name of 'Diamat' as a regrettable incident, there is an attempt, by
taking the point of view of 'true' dialectical materialism to thereby
reverse the situation which has resulted in the theoretical blockage in
modern physics.

Both tendencies reduce dialectical materialism into a useless instrument; it is difficult to believe that problems presented to humanity by
the growth of science in contemporary capitalist society can be reduced
to a debate between 'those who, on the basis of evidence and intuition,
condemn recourse to those new mathematical methods and logic which
physicists and scientists in general seek to use to continually deepen
their understanding of reality' and those who 'judge not merely as
useful but indispensible the use of these methods'.[7]

## THE NATURE OF COMMODITIES

Marxist analysis performs its main task of demystification by developing the conceptual instruments to bring into the open the social characteristics of those properties of things which otherwise appear as objective and natural. The most general case is that of commodities. Marx explains that:

> A commodity is therefore a mysterious thing, simply because in it the social character of men's labour appears to them as an objective character stamped upon the product of that labour; because the relation of the producers to the sum total of their own labour is presented to them as a social relation, existing not between themselves, but between the products of their labour.[8]

This happens in a capitalist society:

> Since the producers do not come into social contact with each other until they exchange their products, the specific social character of each producer's labour does not show itself except in the act of exchange.

> To the latter (the producers), therefore, the relations connecting the labour of one individual with that of the rest appear, not as direct social relations between individuals at work, but as they they really are, material relations between persons and social relations between things.

In particular it happens that commodities are transformed into capital; that is they acquire the property of transforming the means of production into methods by which living labour is utilised to produce new capital. These objects, machines and raw materials in the first place, but later, as we shall see in detail in the pages which follow, non-material goods, such as inventions, patents, know-how, and so on, appear to acquire the mysterious property of directly producing new exchange value. In reality the property that comes to be attributed to these objects belongs to the social relationships which they mediate. The means of production and of life which the capitalist class possesses are in fact the means by which they are able to constrain the class of those who possess only their own labour-power to accept a social relationship under conditions which the capitalist class determines. In this way the constraints which derive from the direct relationship of the subordination of the worker to the capitalist appear as a consequence of the process of work as an objective necessity, not only of machines and of materials, but also of technology and of science. Thus Marx explicitly states:

> In this process the *social* characteristics of labour appear to the worker to have been *capitalised* without him . . . the same, naturally, holds for nature and science (the product of the general historical development in its abstract quintessence) confronting the worker as the *power* of capital, becoming detached from the individual's ability and knowledge. By origin, nature and science may be the product of labour but in appearance they are incorporated into capital as soon as the worker engages in the process of work.[9]

It is this observation that we take as the starting point in attempting to distinguish the 'fetish-like' character (in the precise sense which Marx gave to this term) which science and technology assume in present-day capitalist society.

## INFORMATION AS A COMMODITY

What we have said should clarify our attempt to formulate the concept of 'science' so that we can understand its role in capitalist society as more than merely a productive force. To regard science as a productive force is in fact only one aspect of reality, an aspect which, if it is one-sidedly assumed to represent all of reality, helps attribute to science an objectivity which excludes *a priori* any social conditioning. Because of this we do not attempt here to raise fundamental questions about the concrete mode in which science operates as a productive force in present-day capitalist society. Questions of this type could approach, for example, the mechanisms that mediate between research and economic growth in different countries in the capitalist area, including the effect of their differing relationships with the centres of imperialism, and the interactions of the multinational corporations with diverse sectors of pure and applied research. Neither do we approach the question of the role of the capitalist state in the organisation and the financing of research and of its relationship with private capital.

It seems more important and urgent to us to investigate the nature of the 'social character' which science — at least at this stage of the development of capitalism — acquires.

In our attempt we will extensively utilise the Marxist category of 'commodity'. We are aware of the fact that the role of commodities in mercantile society was very different from that in capitalist society in the two phases analysed by Marx (manufacture and big industry). It is also the case that the significance of the category of commodity is different in the present phase of imperialism, which is characterised by the concentration of vital sectors of production into the hands of the multinational companies, the massive expansion of the service sector, and in general what Marx referred to as 'non-material production'. In the context of our present concerns this is particularly important in respect of the planned production of technological innovations under the control of capital. What we must emphasise is that, despite the lack of a full Marxist analysis of the dynamics of contemporary capitalism we can still use the method of analysis of 'non-material production' as a particular type of commodity with a double nature, of use value and exchange value. Although these are related to one another in a more mediated manner than is the case for the Marxist theory of value as applied to material commodities, in the final analysis abstract labour takes on the character of a commodity labour force. It is clear that this fundamental characterisation of non-material production as a commodity does not exhaust all its properties and functions. Non-material products are not quite the same as an ideal commodity destined for immediate consumption in a competitive system — rather they are commodities in an oligopolistic system whose role it is to enter into the process of production of other commodities. However, we wish only to refer to this conceptual framework here so as to permit us to recognise the existing relationships of production as capitalist, hence making it possible to attempt a scientifically correct — if only schematic — description of contemporary society.

The full growth of capitalist society is characterised by Marx by the fact that 'the entire process of production does not appear as part of the immediate activity of the worker but as a result of the technological employment of science'. None the less the most advanced area of this growth required a qualitatively new condition; this is attained 'only when big industry has already attained a higher level and all the sciences are captured for the service of capital . . . *Now invention* becomes an economic activity and the application of science to immediate production a criterion determined and demanded by production itself.'[10]

Because the production of inventions becomes an economic activity, inventions are a particular form of commodity. This point provides the key for analysing one of the significant aspects of present-day capitalist society, in which the production of non-material goods in the form of commodities has become very important.

It is not only inventions which are produced in the form of commodities, but also an important amount of other *information* relevant to the process of production: know-how, industrial organisation, management — or to that of consumption: marketing, advertising, and so on. All these have come to be produced in capitalist style — that is, to speak in Marxist terms, they have become the result of productive labour (of exchange value).

Furthermore, there has been an enormous growth of information produced as a commodity which becomes directly 'consumed': from mass communication (radio, television, newspapers, magazines, records, tapes, and so on) through individual communication (telephones), to teaching. Marx had already clearly underlined this in relation to teaching.

> If we may take an example from outside the sphere of production of material objects, a schoolmaster is a productive labourer, when, in addition to belabouring the heads of his scholars, he works like a horse to enrich the school proprietor. That the latter has laid out his capital in a teaching factory, instead of a sausage factory, does not alter the relation.[11]

None the less, he adds that 'the great majority of this work is not formally submitted to capital, but serves the needs of transition [towards the mode of capitalist production].'[12]

Successively, above all in Britain and the United States, companies concerned with 'knowledge factories' developed until, as a result on the one hand of the spread of the demand for instruction and on the other because of the interests of capital in a better qualified work-force, general instruction at the lower level, which was not directly profitable, was delegated to the state. None the less many specialised teaching establishments remained in private hands.

The greater part of the sphere of production of information is subject to the capitalist mode of production. That is to say, the proportion of complex capital which has become invested in this sphere of production, with a consequent absorption of salaried workers, has become significant. Today, unlike the situation at the time of Marx, their salary is a capital investment, and not a consumption of income. Their product is in fact destined for the market.

Without examining in detail the capitalist mode of production of information as a commodity, it seems clear that some aspects of the submission of labour to capital in this sphere recall those which occur in the general process of the capitalist production of material goods: the division of labour, with the relative fragmentation and repetitiveness of work, the hierarchisation of function, the alienation of the products of labour from the worker, the way in which capital confronts the worker as if it was a foreign power – in a phrase, the submission of the labour process to the valorisation of capital.

Evidence of the tendency of information to become a commodity is provided by the transfer of technical information – that is the distribution and consumption of the commodity – in the United States.[13] This system is, according to the author, Director of the National Technical Information Service, inadequate to guarantee the transfer to the users of a production of technological information which has grown in volume about sixteen times between 1930 and 1970.

Moreover, the requests for speed in the transfer of information from producers to consumers is increasing: 'competition is a partial cause of the demand for speed, as is a general cultural change that emphasises the value of time'. The inefficiency of the present system depends, apart from various other factors, on the fact that the users are confronted by a price mechanism that bears little or no relation to the satisfaction of their necessities. However, 'in the services offered by commercial enterprises higher prices usually result in better system response and greater user satisfaction'. Because of this, concludes the author, measures to improve the efficiency of transfer are necessary and will take place: 'much greater standardisation of components of the information system and greater reliance on pricing for full cost recovery in order to render higher quality service'. This will result from a distributive system in which the greater capacity for management in the private sector must be accompanied by greater effort on the part of the federal authorities to encourage the integration and co-ordination of its various sectors. It is not necessary to say more about the characteristics of the process of conversion of information to a commodity which this analysis reveals.

None the less we do not wish to imply that *all* the productive processes of information have been absorbed within the sphere of private capital. Parallel with the growth of this process, the power of state intervention into all productive activities, and in particular those of non-material goods, has also grown. When the state intervenes into a productive sector by direct or indirect investment, it does not in any way change the capitalist character of the relationships of production. The fact that an important sector of technological information is produced by state organisations or by the state funding of private ones does not substantially alter the conclusion that information is a particular form of commodity.

Some specific differences none the less exist between the production of information and of material goods; in particular the difficulty in concentrating the production in a site as rigidly controlled as a factory, with a disciplinary regime and surveillance, makes the production of information harder to submit to the process of speed-up, and to constant increases in productivity and growth rate. None the less, productivity in the tertiary sector grows as rapidly as productivity in industry or agriculture.[14]

Furthermore, at first sight information appears to be a very different commodity from others. From the point of view of use value, it may be consumed indifferently by many or by few people without, because of this, anyone having to give up any part, great or small, of what is received. Therefore, to succeed in reducing information to a commodity and thereby attribute an exchange value to it, it is necessary to ensure that it may be used only by those who have 'acquired' it. Hence the legal protection which obliges those entering into possession of certain information to pay the producer a fixed price, or else to provide limitations which physically block those who cannot meet the price from access to the instruments which provide information.

The price of information therefore, seems much less related to the time required to produce it than to the number of consumers. But examining this aspect in more detail, one sees that in the present state of growth of capitalism, which is characterised by a growing differentiation of layers of consumers, the difference between information as a non-material commodity and commodities in the form of material goods is much smaller than was the situation with the production of goods in capitalist society when analysed by Marx. Indeed, one might even say that information becomes a commodity only when accumulation has become mainly directed towards the production of new use values. This stage is characterised by a social mechanism of destruction of use value in advance of the natural processes which render objects — both consumer goods and the means of production themselves — unusable through physical deterioration. In this way the pre-conditions by which information is made quantitative and its consumption measurable (necessary conditions for its transformation into commodity) are created.

Thus, in the same way as for almost all material goods today, there has developed a rapid obsolescence of information and the limitations on the way it can be used, making necessary the systematic and continuously increasing production of new information. From a situation of the free utilisability of information — information which maintains its use value substantially unchanged over time — there is substituted the private consumption of information which is not utilisable unless it is consumed as soon as it is produced. Thus there is created the condition of exhange by which information has become a commodity. Furthermore, one may observe that the mechanism of price formation seems to be identical for material goods and for non-material goods; this is the final requirement which enables us to assert that we have arrived at a situation of the transformation into a commodity of the greater part of the information produced by mature capitalist society.

The substantial uniformity of the forms of commodities, independent of whether they are material or non-material in nature, becomes clearer when one examines those which are means of production. A patent, for example, is a commodity* acquired by a capitalist to be utilised in production in exactly the same way as is machinery.

A new technology — from its birth in a research laboratory to its utilisation in productive relationships — comes above all to have use value for capital. It does not have direct use value for all members of society any more than do machines involved in capitalist production; in order to be able to become a commodity it is necessary in the first place for the technology to be used in the process of valorisation of

---

*Historically the patent has not always been a commodity; at its start it was a way of inhibiting competition in the use of an invention so as to support its inventor, alone or in company with others who could dispose of the necessary capital. This was the case — to take a well-known example — with James Watt. In what we can describe as the 'artisan' phase, the patent came to be a commodity which the inventor, as an autonomous producer, sold to the capitalist who intended to exploit it. Edison is the representative of this phase. Finally in the technological phase of capitalism the patent is a complete commodity, no longer produced by independent workers but by salaried workers; the process of production of innovation is subsumed by that of capital.

capital. The plan for an assembly line, on a par with the machines which physically constitute that same assembly line, is a commodity with a double function: a means of production and an instrument for the production of surplus value.

We are thus dealing with commodities whose use value presupposes generalised commodity production (the reduction of the entire labour force to the status of a commodity); such commodities are utilised to intensify the exploitation of the labour force. The social relations among the producers are reflected in the exchange value of the product of their labour and appear as a social property of this product; in addition at the base of capitalist society, these relationships between capitalist and workers are reflected in the production processes of big industries based on advanced technology. This is not to abandon the distinction between 'productive forces' and 'relationships of production' (concepts which Marx explicitly placed in a dialectical relationship), but neither is it to set up an inpenetrable barrier between natural and social spheres. The growth of productive forces is in reality a process during which historical elements which are derived from social relationships are woven into the objectivity of the relationship between humanity and nature. The productive forces manifest themselves in a concrete form as *productive forces of capital*.

## SCIENCE AS A PRODUCTIVE FORCE AND THE NATURE OF 'PURE' SCIENCE

The Marxist concept of productive forces includes in the first place humans themselves — that is, the workers in a capitalist society. It is therefore a particular socially determined growth of productive forces, characterised by technical innovations and scientific discoveries destined to extend to the maximum the surplus value which can be extracted from the masses, and for the same reason to inhibit that 'full growth of the individual productive forces', which is the sole condition able to guarantee the 'reduction of the labour-time for the entire society to a minimum and to increase the free time of all for their personal growth'.[15] This shows clearly the inadequacy of all theories of the neutrality of technology. It is interesting that even a phenomenological analysis of the correlation between particular consequences of modern technology and its specific function, the valorisation of capital, is sufficient to demolish the thesis of its neutrality. Thus Commoner has noted:

the crucial link between pollution and profits appears to be modern technology, which is both the main source of recent increases in productivity — and therefore of profits — and of recent assaults on the environment. Driven by an inherent tendency to maximise profits, modern private enterprise has seized upon those massive technological innovations that promise to gratify this need, usually unaware that these same innovations are often also instruments of environmental destruction. Nor is this surprising, for, as shown earlier, technologies tend to be designed at present as single-purpose instruments. Apparently, this purpose is, unfortunately, too often dominated by the desire to enhance productivity — and therefore profit.[16]

The concept of neutrality is nothing other than a specific form of fetishism, which attributes an objective intrinsic property to a product of human activity labour which actually derives from the social relationships which intervene between them.

It is usual to distinguish a 'pure science', defined in general as a disinterested activity of investigation of reality, from a science applied to commodity production. Such a dichotomy is integral to the organisation of scientific work. Since this organisation emerged fairly recently, so did the development of the dichotomy in the body of science. Despite the evident historicity of this distinction, we find a frequent characterisation of pure science which is typically vague and lacking any historical specificity which claims that the origins of science lie in some generic spiritual and metahistorical human characteristic. By contrast, we argue that to claim that the base of pure research remains today, as in the past, the innate curiosity of humanity towards the reality which surrounds it, the desire to know and investigate the unknown, human beings' 'natural' capacity to interpret rationally the links that exist among the phenomena which fall on their senses, is either trivial or directly misleading.

'Human nature' can be explained only as a historical process. The curiosity of Galileo is not the same as the curiosity of a present-day physicist who studies elementary particles with a high-energy particle accelerator, because both the social contexts and the functions that they perform are different. As Marx writes: 'Hunger is hunger; but the hunger which is satisfied with cooked meat eaten with a fork and knife is a different kind of hunger from the one that devours raw meat with

the aid of hands, nails and teeth'.[17] If, therefore, we seek to examine
the function which 'pure science' fulfills today, we must distinguish
above all the effective utilisation of results, techniques and methods of
the activity of 'pure' science in other sectors, from the superstructural
role which the production of pure science plays in its specifically
cultural form. Whilst we regard the second aspect as the more important, we should first discuss critically a thesis according to which, the
dominant feature at the level of production is an interest by the producers of advanced technology in the consumption of their goods by
other producers of science rather than an immediate interest by capital
in the growth of science in terms of its possible technological application. The growth of space science is usually cited in this context, as
being an important stimulus to the production of advanced technology.
In reality, however, the argument that pure scientific research represents
above all a form of non-productive consumption of advanced technology necessary to maintain a higher level of demand for technological
goods cannot be sustained. It may be that in particular conditions of
crisis the growth or slackening of investment in scientific research can
be utilised by the capitalist state as a fly-wheel for the economy. This
would be to consider only the fluctuations of money spent around
some sort of middle value, and would not help enable one to understand the importance which scientific research is coming to assume, nor
its connection with technology and production.

On the face of it, it might seem that the interest of capital in the
production of 'pure science' cannot be directly explained by the possibility of the rapid utilisation of the results of this production, because
it has been found that the interval of time between a scientific discovery
and its technical application has not diminished in recent decades, but,
if anything, has increased.[18] This observation, however, is valid only in
relationship to the introduction of a completely new technology
following an important scientific discovery; if one considers instead the
situation in intermediate technology, one finds a steady increasing
rhythm of utilisation. But a further argument which demonstrates a
strict connection between scientific research and economic growth can
be found in the calculations of D. J. de Solla Price.[19] Price demonstrates that the rate of scientific and technological growth in recent
centuries, when measured by a series of variables which the author considers significant (the numbers of universities, of scientific doctorates,
of engineers, of scientific publications, abstracts and important discoveries, of kilowatt hours of electrical energy produced, and so on) has

normally been exponential. The most salient result which emerges from the elaborate graphs* of de Solla Price consists in the fact that the doubling-time of the number of major physicists (twenty years) is equal to the doubling-time of the gross national product. This coincidence strengthens our conclusion that science in our own time is more closely related than in the past to the growth of productivity.

### THE SOCIAL FUNCTION OF 'PURE' SCIENCE

In fact 'pure research' often provides a sensitive test for the efficacy of technological products, making possible the mass development of pretested advanced technology for the production of commodities (for example, computers or miniaturised circuits). In this sense 'pure research' plays an important role in the stimulation of the consumption of goods with a high content of advanced technology in all the leading sectors of the economy. One role of 'pure science' therefore appears today to be that of a method of creating systems through which it is possible to test out technology in the laboratory, a gigantic test-bed for its use in applied science.

Secondly the creation of new languages and methods of science is also important in so far as it provides a starting point for the production of new information commodities. In other words it is not the specific specialised content of the activity of pure research that is utilised in the production of information for the market, so much as the methodologies of production of 'applied' science. It is enough to think of the techniques and methods of computers or operational research, or, in more general terms, the extension of the advanced mathematical methods introduced originally in the study of physics to the practical social ends of organising the productive process.

It is in this sense that a very important function has developed for that restricted stratum of scientific leaders, which, at the head of a mass of middle-ranking scientists[†] signals important changes in the style of various sectors of research and leads the work of the masses towards new models and new experimental techniques.

---

*We wish to make it clear that the author utilises his graphs and the data in order to maintain a thesis with which we do not agree, but despite his thesis, the data themselves are relevant to our present argument.

†We may note that the number of these latter has increased, according to the statistics cited by de Solla Price, more rapidly than has the gross national product. The number of scientific leaders, on the other hand, has grown, he claims, only as fast as the gross national product.

In the third place the big laboratories (as we will see better later on) are ideal places in which to experiment with the introduction of new methods of control and management of a complex integrated productive organisation employing extremely specialised labour of the highest technical level. The substantial validity of this analysis can be seen in a handout from the Public Relations Office of the largest European organisation for fundamental research, CERN (the European Centre for Nuclear Research) at Geneva. In describing the activity of the organisation the handout notes that 'CERN has without doubt directly helped the growth of the technology of production'. In fact the organisation of nuclear instrumentation 'is at the base of progress achieved in the determining sectors of technology', while, as examples of techniques 'developed in collaboration with industry', it cites those of 'very high voltage' (hundreds of thousands of volts), 'short time responses' (hundredths of thousandths of seconds), 'very low pressures approaching that which can be found on the lunar surface', low temperature and superconductors. CERN 'participates in a decisive manner in the growth of systems which are capable of achieving the better development of a group of centralised computers with a great number of terminals and read-in and read-out points'. Finally it is noted that CERN uses experimentally 'the most modern methods of planning the control of all its activities, since it is known that demanding research requires precise technology, and precise technology demands in parallel rigorous management'.

The similarity between the great research centre and the productive structure are recognised explicitly in many official CERN documents. For example, in a report from the labour group concerned with personnel at CERN it is noted that it 'must again be stressed that CERN is predominantly an industrial not an academic employer'.[20]

To sum up. Although 'pure science' cannot be identified directly with the production of commodities, and therefore with an immediate economic activity, such science can none the less be seen to have two distinct functions in capitalist society: in the first place the 'body of scientific knowledge', that is the whole *ensemble* of dominant paradigms in varying fields of research, furnishes the ground-work and provides the support and necessary base from which the production of information can grow; in the second place, scientific production has more and more acquired the function of a test-bed for advanced technology and scientific management, and therefore represents a stimulus to technological advances in commodity production.[21]

Being a social activity which is not directly associated with the production of commodities, 'pure science' is usually seen as the generation of ideas, based on prior ideas; part of an autonomous process which is only accidentally and casually a subject of concern and direction from society. Such concern and direction none the less are not supposed to alter the structure and content of science, but only at most to influence its rhythms and the mode of its growth.

This point of view is common both to those who consider science as a simple reflection in human consciousness of a given objective reality, and also to those who hold a purely positivist view of science as having the aim of simply and economically linking the largest possible number of empirical observations. For the former, the task of science is to bring into the light something which is already given and complete in all its parts and connections. The latter consider that science merely provides the most rational and efficacious method to this same end. In reality these two points of view have in common the property of being 'philosophical', in the sense that they consider science as the result of the mental activity of the abstract intellect of humanity, faced with a nature which is uncontaminated and unchanging in itself. They thereby eliminate the social struggle, economics and history as irrelevant to the gnosological problem, which is regarded as eternal and unchangeable.

It is not necessary to emphasise how little such an interpretation has to do with Marxism, even though such a 'philosophy of nature' often appears dressed up in the style of dialectical materialism. Instead we must try to understand how the production relationships of advanced capitalist society influence the models which members of this society may make of their own relationships with nature, and how the models of relationships with nature are reflected in models of social relations. In the first case we must trace, in scientific production, the reflection of the mode of material production, and in the second to identify those contributions to the production of ideology that are linked to the process of scientific production. There is a celebrated note by Marx in the first volume of *Capital* which serves well to illustrate these objectives:

Technology discloses man's mode of dealing with Nature, the process of production by which he sustains his life, and thereby also lays bare the mode of formation of his social relations and of the mental conceptions that flow from them. Every history of religion even, that fails to take account of this material basis, is uncritical. It is, in reality, much easier to discover by analysis the earthly core of the misty creations of religion, than conversely, it is, to develop from the actual relations of life the corresponding heavenly form of these relationships. The latter method is the only materialistic and therefore the only scientific one. The weak points in the abstract materialism of natural science, a materialism that excludes history and its process, are at once evident from the abstract and ideological conceptions of its spokesmen, whenever they venture beyond the bounds of their own speciality.[22]


## THE PRODUCTION OF 'PURE' SCIENCE – THE BIG LABORATORIES

In discussing the relationships between the dominant mode of production at the material level and the production of science, we should note what Marx stated in general concerning the relationships between different branches of production.

> Under all forms of society there is a certain industry which predominates over all the rest and whose condition therefore determines the rank and influence of all the rest. . . . . It is the universal light with which all the other colours are tinged and are modified through its peculiarity.[23]

In the light of this premise we wish to argue that the production of information as a commodity also dominates the interior of the sphere of scientific production – even that which is not commodity-production – and characterises both the relationship among producers and the relationships between them and the product of their labour. We have shown that the production of pure science represents the necessary base for the production of information, and that the production of information has assumed in advanced capitalist society the characteristics of the capitalist production of material goods, i.e., commodities. Because pure science has come to play a role in the production of technology and information at the industrial base, it can no longer be produced in the artisan way which had distinguished it up to the 1939 – 1945 war. Today there exist the big laboratories* – national and

---

*The material conditions necessary for the existence of big laboratories and big monopoly capitalism, and therefore the massive concentration of capital, all derive from similar sources.

international — which have in the last decades become the centres of scientific research, and which have relegated to a secondary role those small laboratories which are decentralised in individual universities. The big laboratory serves as a multiplier of the productive efficiency of science (concentration of scientists, greater rapidity of exchange of information, the possibility for a group of researchers to use varied equipment, the possibility of simultaneous utilisation of plant by several research groups, and so on) and therefore guarantees that the production of pure science can take the strain of the industrial production of information.

Although historically the prototype of the big laboratory (the Manhattan project) was born prior to the development of the production of information as the dominant mode in science, the justification and consolidation of the big laboratories in scientific research was based on the fact that in the years which followed the 1939 – 1945 war applied science and information in particular had become commodities.

To clarify the connections that exist between the big laboratories and the production of information as a commodity, we wish now to examine in more detail the specific characteristics which the production of science assumes within the big laboratories.

The most advanced methods of organisation management are utilized to maximize the 'productivity' of research. For Harvey Brooks, one of the high priests of American science politics, 'The first question is how to organise staff and direct the search for knowledge so as to obtain the greatest rate of scientific progress for a given investment of human and material resources.'[24] It is not difficult to detect the consequences of this tendency.

First, the organisation of research work tends to become independent of the aims of the research itself, but is determined primarily by the instrumentation used. Since this has come to resemble closely that which is used in the production of technology, in the last analysis it is the capitalistic use of the instruments of labour, that tends to result in the division of labour and the organisation of the production of science. Specialisation has also fragmented labour into simplified components; in each research group the diverse components are directed to different goals, thus creating a separation between researchers and technicians and the establishment of hierarchical relationships within the research group itself. Where in the past the technique to be used in research has been subordinated to the nature of the problems to be resolved, today, following the increase in the number of those researchers whose work is determined by the technical conditions available, the position has been reversed; the choice of the problem is subordinated to the technique.

Thus, according to Yaes,[25] in an open-minded description of the research conditions in the sector of elementary particle physics, 'like the rest of the work force, hundreds of physicists are now being subjected to a lifetime of boring, meaningless, alienated work'. In these conditions, the members of the physics establishment assume, in confronting their colleagues, the same attitudes that managers do in industry, who 'have traditionally regarded their employees, more as a means of production than as human beings'.

Second, as with the production of commodities, time becomes a determining element in the production of a useful research result. A piece of research loses all value if it generates a result which is already known. Research becomes a ruthless race to arrive first, and there is a tendency to reject long-term research to arrive rapidly at a tangible result.

There follows an extremely rapid obsolescence of information, in complete harmony with the general characteristics of all commodities at the present stage of capitalism. This process results in the stratification of consumption, between a restricted elite which disposes of information rapidly and at first hand in a way most likely to achieve the most innovative technical growth, and a mass of consumers for whom the product arrives in a way which is only suitable for use for routine production. Unlike the situation for most research, this subdivision coincides, as is demonstrated in an illuminating study by Morandi, Napoli and Ratto,[26] * with the division of the 'second world' according to the geography of imperialism — the metropolitan producers of advanced technology; the satellite areas, producers of mature technology; and the underdeveloped areas.

Third, in order to determine productive efficiency a quantitative criterion has been adopted as a recognised social measure of success.

---

*In their analysis the authors arrive at the following conclusion concerning 'capitalism and export imperialism' which are characterised by extreme scientificity and precision in their mode of production, representing:
(1) A rigid division of international labour which reflects the divisions of the world into areas which produce advanced technology, mature technology, and a third underdeveloped area; and
(2) A mode of doing science within each particular area which has as its base the maximisation of productivity (bringing in its train, for example, the progressive division of labour between theoretical and experimental) and, in its characterisa-

This is a consequence of the necessity of justifying the productivity of investment in the means of production and wages in relation to invest-ment in industrial sectors of the economy. Thus, the production of 'pure science' takes on the rhythm of the industrial production of information, which, as we have already made clear, represents an indispensible condition for the growth of information as a commodity. It does not matter whether the information is useful; it matters that it can be produced. In the field of high-energy physics alone a hundred pre-prints of work are received in the library of one big laboratory every week, or some 5000 each year. No one is in a position to digest this great mass of information or to select those contributions which are most useful to a particular end. It is necessary to rely on a different criterion of 'value' in order to choose. This is provided by productivity. The necessity of establishing a quantitative criterion by which scientific production can be measured results in a scale which selects those con-tributions which promise to result in the largest number of further publications. The measure of the success of any published work is determined by the number of citations, and the measure of efficiency of an institution is the number of publications it produces. Cole and Cole[27] have used the criterion of the number of citations as an abso-lute measure of the value of an article. As Yaes puts it:

> this technique will be particularly attractive to science administra-tors for several reasons. It sounds plausible and, because it does not involve any subjective evaluation on the part of the administrator in question it also seems objective. Besides, counting citations is easier and less time-consuming than subjective evaluation which, in any event, assumes a degree of technical sophistication on the part of the manager that is equal to that of the scientist he manages. Most important of all this technique will tend to prove what they want it to prove. Scientists who are already famous will have their work

---

tion of scientific labour more and more as exchange value rather than as use value, lays emphasis not so much on the abstract progress of scientific knowledge as on its novel character and therefore of the marketability of the commodity. In this framework, too, as we have recorded, certain techniques come to be develo-ped only when they are innovative, and are abandoned when they arrive at maturity. (That is, they are abandoned in their growth phase and before they have produced appreciable results.) It is very interesting that this conclusion emerges from close examination of a homogeneous field of about 300 works published in a very specialised area of solid-state physics, thus furnishing an 'experiment' which directly supports our general analysis here.

cited more often *precisely* because they are more visible and people pay more attention to what they say.[28]

Young researchers are forced, if they do not wish to be rapidly driven out, to publish a lot of fashionable material without time for reflection. The scientific institutions (centres, laboratories, research groups) tend to concentrate equipment and effort in directions which can secure results according to canons which are accepted and defined by the establishment; a typical example is the construction of ever more powerful particle accelerators.

However, this process is assisted by the same process of concentration that characterises the great capitalist firms. Research becomes concentrated in heavily financed centres of research, eliminating the small laboratories. The National Accelerator Laboratory at Batavia in the United States, which cost $250,000,000 to build, now costs $60,000,000 each year to run. This is despite — or better precisely because of — the fact that the scientific results are minimal. As *Science* puts it: 'In the absence of any new striking discoveries however, more quantitative measurements are planned.'

These developments are not limited to physics, which merely represents a paradigm case. The testimony of the biologist S. E. Luria is indicative in this respect:

> The pursuit of scientific research varies depending on external circumstances, with regard not only to the contents of research but to the way it is carried out — its style. The charming snootiness of the physicists as intellectuals, for example, did not survive the pressure to associate with the military crowd during the 1940's. . . . A medium-big scale, not quite that of physics, but relatively substantial all the same, has overtaken biology. . . . . But the entrepreneurial system does lend itself to opportunism. . . . . A subtle change in ethical standards follows: not necessarily a loss of integrity, but a shift of responsibility from the scholar to the entrepreneur. One sees signs of such a change taking place in biology, in which substantial research support dates only from two decades ago. For example, if someone published some good work, other scientists used to allow him to develop it alone at least for a few years. Now eager researchers rush back from professional meetings to perform the obvious experiments that a speaker had not yet had time to do. Nothing strictly unethical, of course — not according to the ethics of competitive enterprise.[29]

## THE IDEOLOGY OF SCIENCE

Besides the mechanisms which operate directly on the organisation and division of labour within the big laboratories, on the form of product and on its social value, there exist more mediated relations at the superstructural level between the social relations of production which are dominant in society and the form which science assumes as a social product. Above all, because the scientific elite forms part of the ruling stratum through its education, social contacts, and concrete interests of collaboration with the productive structure, the education structures and the mass media, it transmits dominant values and behaviours to the interior of the scientific corporation.

In particular its submission to the norms dictated by the ruling classes is assured by a tight network of consultative organs at all institutional levels. A recent discussion of the scientifico-technical consultative apparatus in the United States[30] shows how this net not merely brings into positions of power the most important group of scientists and engineers, but directly influences the behaviour of the enormous majority of members of the scientific community. Belonging to the apparatus of consultation in fact furnishes scientific prestige and professional advantages which represent an incentive for younger and less well-known scientists. Moreover, affirms the author: 'it is commonly held that the members of this consultative body are the best qualified professional experts, while those who are outside their numbers are considered dilettantes'. From all this comes a very cautious attitude of scientists in the face of the Administration, and the tendency 'not to oppose its scientific and technological policy with excessive vigour or too publicly'.

To point briefly to a second aspect of the link at the superstructural level between the production of commodities and the production of science, it has been noted how in physics the vigorously reductionist modes of explanation used in the first decade of the century have vanished with the development of particle physics. This transformation consists of replacing reductionism by a global description disengaged from the traditional concept of 'dynamic evolution' This transformation may correspond with the passage from mechanisation to automation in the productive process for material goods, namely with the transformation from the use of systems in which the behaviour of a single component determines the behaviour of the resultant system, to a situation in which the over-all behaviour of a complex system is determined by reciprocal feedback in all its components. In other words, it appears to show an example of the relationships which link work practice, by means of which humans intervene actively in nature to transform it, and the conceptual instruments which they use in order to understand nature in the course of their conscious activity. The change which has been described represents a real redefinition of the concept of the scientific explanation of a determinate process, and as a consequence — as Jona-Lasinio has written[31] — 'there no longer exist criteria of truth in the strict sense. In this condition the average theoretical physicist is merely a functionary, creative in the best cases. Theoretical physics can no longer explain anything.'

In order that scientific research can perform the two functions previously discussed (the indispensible support of the production of information and technological test-bed), there is a need for a minority of scientists, mainly gathered together at the highest point of the hierarchical ladder of the scientific community, to have the privilege of establishing the programmes and determine the paradigms within which the routine functionaries, whose role we have just discussed, complete the pattern. One may note in this context the view of Kuhn[32] that the activity of the scientific community in periods of normal — as opposed to revolutionary — science centres around the articulation of the accepted paradigm and the resolution of puzzles within the interior of the paradigm itself. An important difference between Kuhn's view and ours is that we believe that in the production of science, in advanced capitalism, there can coexist different paradigms without a period of crisis (in the sense used by Kuhn) within the scientific community. Such production may be articulated into research programmes which are different but not mutually exclusive and therefore substantially equivalent from the point of view of their knowledge content.

The process of scientific production — the production of a good which, if not immediately a commodity, is none the less socially useful — has become steadily more submissive to the capitalist mode of production of information in the form of a commodity. As a consequence its products are increasingly characterised by the social character which appears to be intrinsic and objective but in reality, reflects the relationships of production.

As in the production of material commodities, the material concreteness of goods which are produced, and of equipment which is developed to produce them — their use value — combines as material and objective support 'with the specific social character which at each given phase of historical growth they possess', so in the production of science the real objectivity of the relationships between humanity and nature are combined with the specific social characters conferred on this relationship by the dominant mode of production. In other words, precisely because it represents objective relationships between humanity and nature, the science which is produced in advanced capitalist society is able to provide scales of values, models of behaviour, forms of organisation, social goals, all of which appear objective and natural; and it is exactly this ideological content which contributes in large measure to what has come to be considered contemporary 'scientific culture'. Let us briefly examine some salient aspects.

First, science furnishes a model of growth based on production as an end in itself. In fact it comes to boast of not being in any way finalised. Science's own 'purity' rejects every 'instrumentalisation'. This 'purity' comes to be ennobled and represented as an autonomous and intrinsic value of science, although is actually no more than the characteristic law of capitalist production according to which 'the scale of production does not depend on given needs but on the contrary the amount that is produced depends on the scale of production (always growing) determined by mode of production itself'.[33]

Second, science presents itself as a body of knowledge which is closed to those who are not trained in its work. As a consequence, no social control over its role and purposes is admitted. This leads to a social model in which those who are competent in a given sphere of activity form a separate body which places itself above common people. It is clear that this form of organisation appears as a necessary condition for the proper mode of functioning of science and therefore as an objectively valid model for all the other institutions of society. It follows that because only the recognised institutions have the power to define who is competent, those who wish to contest the recognised institutions are, by definition *incompetent*, and ought to be put to one side. But it is also clear that the opposite is true; it is the form of the institutions of capitalist society whose origin derives from their function of reproducing the current social relations and which also characterise science as an institution of this form of society.

Third, science is presented as pure objectivity. There follows a model of society in which relationships between humans are determined by objective laws, a society where common people ought to accept that their life is regulated by a 'scientific' organisation of work, that their capacity can be evaluated by 'scientific' means, and in which their place in society can be fixed by an 'objective' scale of values.

Finally, science in its ever more rigorous specialisation shows the path of success. There follows a model of society in which everyone ought to be bound exclusively to an ever more restricted sector of activity, renouncing all participation in the collective life and delegating to the mechanism of the system the solution of social problems.

It is unnecessary to emphasise how this 'suitable model of society' is merely the image reflected, 'as in a mirror', of advanced capitalist society.

ACKNOWLEDGEMENTS

We wish to thank Federico Marchetti for having contributed to the discussion of some of the themes which underlay this work in its initial phases. We are grateful to all the friends who have contributed to the final version of the work with suggestions and discussions. Among them we wish to mention explicitly Esther Damascelli, Alberto Gaiano, Marco Lippi, Bruno Morandi and Franz Navach. We also wish to thank all of those who have critically discussed the ideas presented in this work in seminars at Lecce, Palermo, Naples, Bologna and Rome.

# 4

# On the Class Character
# of Science and Scientists

*André Gorz*

All the talk about the proletarianisation of scientific workers just demonstrates one fact: most scientific workers still do not feel part of the working class. If they did feel part of it they would not discuss their proletarianisation. Do we discuss the proletarianisation of chemical workers, or engineering workers, or electricians, printers, service workers? We do not.

Why then do we discuss the proletarianisation of scientific workers? For a quite simple reason: our minds still are not quite reconciled with the fact that the words *scientific* and proletarian fit together. In the back of our minds, we still find it hard to believe — or even outright shocking — that a person with a degree in science should be considered a worker just like a person with a 'degree' in plumbing, drawing, tool-making or nursing.

To most of us, whatever the political convictions we profess, there still is an essential difference between a scientific worker and, for instance, a metal worker: the adjective *scientific* does not, in our subconscious, refer to a *skill,* a *craft* or an *expertise* like any other; it refers to a *status*, to a position in society. And we speak of proletarianisation *not* because we feel and think like proletarians, but only because our previous status is threatened, because our privileged position is now in jeopardy.

If we had the courage to be quite frank, most of us would admit that proletarianisation was a shock to them; they had expected their training in science to earn them an interesting, well-paid, safe and *respected* position. They felt entitled to it. And they felt entitled to it because most of them were brought up in the traditional belief that knowledge is the privilege of the ruling class and that the holders of knowledge are entitled to exercise some power, to hold some privilege. If we are quite frank, we shall admit that most of us had, or still have, an inherently elitist view of science; a view according to which, *those who know* are a small minority and must remain a minority. Why must they? Because science *as we know it* is accessible only to an elite: everyone can't be a scientist or have a scientific training. *This is what we have learned at school.* Our whole education has been devoted to teaching us that science cannot be within the reach of all, and that those who are able to learn are superior to the others. Our reluctance to consider ourselves as being just another type of worker rests upon this basic postulate: science is a superior kind of expertise, accessible only to a few.

This is precisely the postulate which we must try and challenge. Indeed we must ask: why has a science — or systematised knowledge generally — been so far the preserve of a minority? I suggest the following answer: because science has been shaped and developed by the ruling class and for its benefit in such a way as to be compatible with its domination — that is, in a way that permits the reproduction and the strengthening of its domination. In other words, our science bears the imprint of bourgeois ideology and we have a bourgeois idea of science.

With these remarks, I do not intend to indulge in extreme and primitivistic simplifications. I do not mean to say that science itself is something bourgeois and that we have to discard all the special knowledge and expertise we may hold, considering it an undue privilege and a result of bourgeois education. When saying that our idea of science and our way of practising it are bourgeois, I rather have in mind the following three aspects:

(1) the definition of the realm and nature of science;

(2) the language and objects of science; and

(3) the implicit ideological content of science.

(1) As regards the definition of what is and what is not scientific, our society has a quite peculiar view: it calls *scientific* the knowledge and skill that can be systematised and incorporated into the *academic* culture of the ruling class; and it calls *unscientific* the knowledge and skill that belong to a popular culture which, incidentally, is dying out rapidly. Take a few striking examples:

(a) In medicine, in France (amongst other bourgeois countries), allopathy, relying on heavy synthetic drugs, is considered scientific, whereas homeopathy, acupuncture and plant medicine, all of which spring from ancient popular culture, are considered unscientific and are condemned by the medical profession.

(b) When the research department of a large automobile firm puts a new engine on the market, this engine, of course, is the produce of scientific expertise. But when a group of amateurs or craftsmen who have never been to university build an even better engine, using hand-made parts, this, of course, is something unscientific.

(c) When experts in industrial psychology organise the work process so as to divide the workers and to make them work to the limits of their physical capabilities, this is something scientific. But when workers find a way of uniting, striking and of reorganising the work process so as to make it as pleasant as possible, this, of course, is something unscientific.

What are the criteria behind these distinctions? Why are homeopathy or plant medicine skills and allopathy 'science'? Why do we call the invention of a mechanic or a toolmaker the product of *craftsmanship*, and the same invention the product of *science* and *technology* when it is presented by an engineering firm? Why is the management psychologist a 'scientific expert' and shop stewards or militants nothing of the sort when they expertly turn the tables on the expert? Why does one speak of 'the scientist as worker' and never of 'the worker as scientist'?

The answer, I suggest, is that our society denies the label of 'science' and of 'scientific' to those skills, crafts and knowledge which are not integrated into the capitalist relations of production, are of no value and use to capitalism, and therefore are not formally taught within the institutional system of education. Therefore, these skills and knowledge, though they may rest on extensive studying, are not included in the dominant culture. They have no status within this culture; they are not institutionally recognised 'professions' and they often have little or no market value — they can be learned by anyone who cares to from anyone who cares to be learned from. Our society, however, calls 'scientific' only those notions and skills that are transmitted through a formal process of schooling and carry the sanction of a diploma conferred by an institution. Skills that are self-taught or acquired through apprenticeship are labelled 'unscientific' even when, for all practical purposes, they carry as much efficiency and learning as institutionally taught skills. And when we look for an explanation of this situation, the only one we shall find is a social one: self-acquired knowledge, however effective, does not fit into the pattern of the dominant culture; and does not fit into it *because it does not fit into the hierarchical division of labour* that is characteristic of capitalism.

Just suppose for a moment that a boilermaker or a toolmaker in a factory were credited with as much expertise as a university-trained engineer: the latter's authority, and thereby the hierarchical structure, would be in jeopardy. Hierarchy in production and society over all can be preserved only if expertise is made the preserve, the privilege, the monopoly of those who are *socially selected* to hold both knowledge *and authority*. This social selection is performed through the schooling system; the main — though hidden — function of the school has been to restrict the access to knowledge to those who are socially qualified to exercise authority. If you are unwilling or unable to hold authority, you either will be denied access to knowledge or else your knowledge will not be rewarded by any existing institution.

To sum up. In our society, the nexus between authority and knowledge is the inverse of what it is supposed to be: authority does not depend on expertise; on the contrary, expertise is made to depend upon authority: 'the boss can't be wrong'.

(2) This social selection of the knowledgeable and the expert is performed mainly through the way in which scientific knowledge and expertise are taught. The method of teaching and, beyond, the whole schooling programme are devised in a way which makes science inaccessible to all but a privileged minority. And this inaccessibility is not due to some intrinsic difficulty of scientific thinking; it is rather due to the fact that in science — as in the rest of the dominant culture — the development of theory has been divorced from practice and from ordinary peoples' lives, needs and occupations. We may even say that science was *defined* socially as being *only* that kind of systematised knowledge which has no relevance to the daily needs, feelings and activities of people.

Modern science was initially conceived of as being impermeable and indifferent to human concerns, and concerned only with dominating nature. It was not intended to serve the mass of the people in their daily struggle; it was meant primarily to serve the ascending bourgeoisie in its puritan effort at domination and accumulation. The ethics and ideology of the puritan ruling class clearly shaped the ideology of science, generating the notion that the scientist must be as self-denying, insensitive and inhuman as the capitalist entrepreneur.

In this sense, there has never been anything like 'free' or 'independent' science. Modern science was born within the framework of bourgeois culture; it never had a chance to become popular science or science for the people. It was confiscated and monopolised by the bourgeoisie, and scientists, like artists, could only be a dominated fraction of the ruling class. They could enter into conflict with the rest of their class but they could not break out of bourgeois culture. Nor could they go over to working class; they were — and still are — separated from the working class by a cultural gulf.

This gulf is reflected in the semantic divorce of expert from everyday language. The semantic barrier between scientists and ordinary people must be seen as a class barrier. It points to the fact that the modern development of science — like that of modern art — was culturally cut off, from the outset, from the over-all culture of the people. Capitalism has to an unprecedented extent sharpened the division between practice and theory, manual and intellectual labour; it has created an unprecedented gulf between professional expertise and popular culture.

During the last decades, it has achieved something even more astounding; needing huger and huger amounts of scientific and technical expertise, it has cut this expertise into such minute fragments and so many narrow specialisations that they are of little if any use to the 'experts' in their daily lives. In other words, to the traditionally bourgeois scientific culture has now been added a new type of technical and scientific subculture that can be used *only* when combined with other subcultures in large industrialised institutions. The holders of this specialised expertise are professionally as helpless and dependent as unskilled or semi-skilled workers. The kind of scientific expertise which most people are taught nowadays is not only divorced from popular culture, but it is impossible to integrate into any culture; it is culturally sterile or even destructive.

We here reach the central aspect of the class nature of modern science: whether theoretical or technical, comprehensive or specialised, so-called 'scientific' knowledge and training are irrelevant to peoples' lives. There has been a tremendous increase in the quantity of knowledge and information available to us; each of us, and all of us together, know a great deal more than in previous times. Yet this enormously increased quantity of knowledge does not give us a greater autonomy, independence, freedom or capacity for solving the problems we meet. On the contrary; our expanded knowledge is of no use to us if we want to take our collective and individual lives into our hands. The type of knowledge we hold is of no help to us in controlling and managing for ourselves the life of our community, city, region or even household.

The expansion of knowledge rather has gone in parallel with a diminution of the power and autonomy of communities and individuals. In this respect, we may speak of the schizophrenic character of our culture: the more we learn, the more we become helpless, estranged, from ourselves and the surrounding world. This knowledge we are fed is so broken up as to keep us in check and under control rather than to enable us to exercise control. Society controls us by the knowledge it teaches us, since it does not teach us what we need to know to control and to shape society.

(3) This brings us to the third aspect of the class character of modern science: the ideology that underlies the solutions it offers. Science is not only functional to capitalist society by way of domination through the division of labour which is reflected in the language, definition and division of its disciplines. It is also functional to capitalism in its way of asking certain questions rather than others, of *not* raising issues to which the system has no solutions. This is particularly true in the field of the so-called human sciences, including medicine; they devote much effort to find ways of curing and checking illnesses and dissatisfactions; they devote much less effort to finding ways of preventing illnesses and dissatisfactions; and they devote no effort at all to finding ways of dispensing with all the experts in health and satisfaction, although the only sound solution would be just this; to enable all of us — or at least all those who wish to — to cure the common diseases, to shape housing, living and working conditions according to our needs and desires, to divide labour in a way we find to be self-fulfilling and to produce the things we feel to be convenient and pretty.

Western science, as it is constituted presently, is inadequate to all these tasks. It does not offer us the intellectual and material tools to exercise self-determination, self-administration, self-rule in any field. It is an expert science, monopolised by the professionals and estranged from the people. And this situation after all is not surprising; Western science was never intended for the people. Its main relevance, from the outset, was to machinery that was to dominate workers, not to make them free.

What makes the situation so complicated is the fact that intellectual workers are both the beneficiaries and the victims of the class nature of Western science and of the social division of labour that is built into it.

Whether we like it or not, we are beneficiaries of the system since we still hold significant though dwindling privileges over the rest of the working class. Manual, technical and service workers rightly consider that scientific workers belong to the ruling class; as carriers of bourgeois culture, the latter are bourgeois, at least culturally. The scientific workers in the manufacturing and mining industries may be considered to be bourgeois socially as well. In France, for example, the engineers of the state-owned coalmines are one of the most reactionary and oppressive groups of the French bourgeoisie. In most factories, production engineers as well as management experts are distrusted and hated by the workers as their most immediate enemies; not only do they hold significant privileges as regards income, housing, working conditions – it is these technical and scientific experts, too, who engineer the oppressive order of the factory and the hierarchical regimentation of the labour force.

Clearly, it must be seen that the class character of the capitalist division of labour and the class conflict between production workers and scientific and technical personnel will not disappear from the factory floor through mere public ownership of the industries. Public ownership will not destroy class barriers and antagonisms, even if it were accompanied by extensive wage equalisation and change of attitudes. Class distinctions in the factories will disappear only with the disappearance of the hierarchic capitalist division of labour itself, a division which robs the worker of all control over the process of production and concentrates control in the hands of a small number of employees. The fact that these employees – whom Marx called the officers and petty officers of production – are themselves part of the 'total workers' (*Gesamtarbeiter*) is quite irrelevant as regards their class position; they are in fact paid to perform the capitalist's function, which can no longer be performed by one boss and owner. And the job they perform for a salary is in fact perceived by the workers as being instrumental to their exploitation and oppression.

This oppression will persist, regardless of who owns the factory, as long as the technical, scientific and administrative skills requested by the process of production are monopolised by a minority of professionals who leave all the manual tasks and all the dirty work to the workers. Whatever the political views of these professionals, they, in their roles, embody the dichotomy between intellectual and manual work, conception and execution; they are the pillars of a system which robs the mass of the workers of their control over the production process, and which embodies the function of control in a small number of technicians who become the instruments of the manual workers' domination.

You may argue, of course, that the technical staff in factories are themselves oppressed, that they too are victims and not only instruments of the capitalist division of labour. This is perfectly true. But I must insist on the point that being oppressed is not an excuse for oppressing others and that oppressed oppressors are in no way less oppressive. Moreover, while engineering and supervisory personnel doubtlessly are oppressed or exploited, they are not oppressed *by the workers whom they are dominating* and cannot expect the latter's sympathy.

I insist on this point because there can be no unity and no common struggle of the various sectors of the working class as long as those workers who hold scientific and technical knowledge and skills do not recognise that they in fact have an oppressive role *vis-à-vis* the manual workers. There is a significant proportion of highly skilled personnel who believe that they must be anti-capitalist and socialist because they are in favour of self-management, that is, in favour of running the plants themselves without being controlled by the owners. In truth, there is nothing socialist in this technocratic attitude; doing away with the owners and their control would not abolish the hierarchic structure of the plant, or laboratory, or administration; it only might alleviate the oppression suffered by employees in responsible positions, without diminishing the oppression these employees inflict on production workers.

All those who want to ignore the class nature of the present division of labour, and the class division between intellectual and manual workers, are in fact incapable of envisaging a classless society and of fighting for it. All they envisage is a technocratic society that may be branded state-capitalist, or 'state-socialist' if you wish, and in which the basic relations of production of capitalism will prevail (as indeed they prevail in Eastern Europe and the Soviet Union).

When saying that intellectual workers are in fact privileged, and objectively in an oppressive role, I do not infer that, in order to be socialists, they must renounce any specific demands and serve working-class interests with guilty selflessness. On the contrary, I am convinced that the abolition of the capitalist division of labour is in the intellectual workers' own interest, because they are victimised and oppressed by it as much as the rest of the working class.

The scientific workers' proletarianisation began some ninety years ago in Germany, when Carl Duisberg, who was research director at Bayer, submitted research work to the same division of labour as production work. This industrialisation of research has since become universal. As industry found that science could be a force of production, the production of scientific knowledge has been submitted to the same hierarchical division and fragmentation of tasks as the production of any other commodity. The subordination of the laboratory technician or anonymous researcher to his or her boss, and of the latter to the head of the research department, is not very different, in most cases, from the subordination of the assembly-line worker to the foreman, and of the foreman to the production engineer, and so forth. The industrialisation of research has been responsible for the extreme specialisation and fragmentation of scientific work. The process and the scope of research have thereby become as opaque as the process of production, and the scientist has in most cases become a mere technician performing routine and repetitive work. This situation has opened the way for the increasingly military uses of scientific work, and the latter, in turn, has led to a further hierarchisation and specialisation of research jobs. Science is not only militarised as regards its uses and orientations; military discipline has invaded the research centres themselves as it has the factories and administrations.

In short, scientific work has undergone much the same process as production work from the early nineteenth century onward; in order to control and to discipline production workers, the early capitalist bosses have fragmented the work process in such a way as to make each worker's work useless and valueless unless it is combined with the work of all others. The boss' function was to combine the labour he had first fragmented, and the monopoly of this function was the base of his power; it was the pre-condition for separating the workers from the means of production, and from the product. In the production of science, control and domination of the scientific labour force are even more vital than in other areas of commodity production; should the production of knowledge escape the control of the ruling class, the holders and producers of knowledge might take power into their own hands and establish a more or less benevolent or tyrannical type of technocracy. The bourgeoisie has been persistently haunted by this danger since the second half of the nineteenth century. To make their power safe, capitalists had to make sure that knowledge could not wield autonomous power but would be channelled into uses compatible with or profitable for capital.

There are of course two obvious ways to bring science — and knowledge generally — into the power of the capital-owning class.


(i) The first way, which is widely practised in the universities, is the socio-political selection — and promotion — of scientists.

Scientists in responsible positions must belong to the bourgeoisie and share its ideology. During and after their schooling process, appropriate steps are taken to persuade the ambitious that their interest lies in playing the establishment's game. In other words, scientists tend to be bought off, to be co-opted into the system. They will be given positions of power and privilege, provided they identify with the established institutions. And their power, which is both administrative and intellectual, has a definitely feudal aspect; the big bosses of medicine or science departments in the universities hold the discretionary powers of the feudal landlord in previous times. The hierarchy in the production of science is as oppressive as in factory production. The big bosses of science must be seen as watchdogs of the bourgeoisie, whose particular function it is to keep the teaching, the nature and the orientations of science within the bounds of the system.

The domination of these bourgeois scientists over science would be impossible, of course, without the consent of those over whom they rule. As usual, two instruments are used to manipulate young scientists into submission to the bosses: ideology and competition.

There is not much point enlarging on the current ideology of science, one that pretends to be value-free and which, under the pretence that science has no other purpose than to accumulate knowledge, accumulates any kind of knowledge, namely, 90 per cent useless knowledge and 10 per cent that is useful to the system. The important point is, unless he or she accepts the prevailing ideology, a young scientist will not get far, will not make a career, but be eliminated by the institution.

Such elimination is made possible by the vast abundance of candidates seeking to do research work. The bosses of science, and through them the system, base their domination on the tremendous surplus of students that can be found in all industrialised societies. This surplus of students enables the bosses to organise the rat race. In other words, the potential surplus of scientific labour has the same effect as the reserve army in industrial labour; it strengthens the boss *vis-a-vis* the workers and enables him to play them off against each other.

But competition between researchers has an even more important consequence; it leads to the most extreme forms of specialisation. The reason for this is obvious; to make a career, a research scientist must produce something original. This can best be done by pushing research into the most hair-splitting details of an otherwise trivial field, the aim of academic research being not to produce some knowledge relevant to a definite issue, but only to prove the researcher's capability: a 'value-free' and 'neutral' capability.

(ii) The extreme specialisation of competing scientists is precisely what capital needs to make its own domination safe. Competing, over-specialised and hair-splitting scientific workers are not likely to unite and to translate knowledge into power. Furthermore, the over-abundance of scientific talent enables the capitalist class to pick those people who seem fittest to serve the interests of the system. This situation also enables the bourgeoisie to stiffen the division of labour in scientific work, so as to keep control over the production of science and to prevent scientific communities from pooling their knowledge and becoming a major force in their own right.

All the modernistic talk about the scientific workers being destined to win power themselves within society because — so the story goes — knowledge and power cannot be indefinitely separated, all this talk is pure rubbish. The scientific workers are in no position to claim or to conquer power because so far they have been incapable of uniting on a class basis, of evolving a unity of purpose and presenting a vision encompassing the whole of society. And this inability is nothing accidental; it merely shows that the type of knowledge held by scientific workers, individually and collectively, is a *subordinate* knowledge, that is a type of knowledge that cannot be turned against the bourgeoisie because it inherently bears the imprint of the social division of labour, of the capitalist relations of production and of capitalist power politics.

The *immediate* interests of scientific workers therefore are no more revolutionary or antagonistic to the system than the *immediate* interests of any other privileged segment of the working class. Quite the contrary; the present specialisations of a majority of scientific and technical workers would be totally useless in a socialist society. The fact that large numbers of scientific and technical workers are unemployed or underemployed, as of now, under capitalism, does *not* mean that a socialist society would have to or would be able to employ them in their present specialisations. People with a scientific or technical training are *not* victims of capitalism because they cannot find creative *jobs* — or any kind of a job — in their present capacity; *they are victims of capitalism because they have been trained in the first place in specialisations that make them incapable of producing their livelihood, and are useless in this and any other type of society.* And they have been so trained for three reasons:

(*a*) to hide the fact that their labour is not needed by the system, that is, that they are structurally unemployed and unemployable;

(*b*) because it would be dangerous not let them hope that through studying they can win a skilled and rewarding job; and

(*c*) because a reserve army of intellectual labour performs a useful function under capitalism.

Therefore, the first step towards the political radicalisation of intellectual labour is not to ask for more and for better jobs, mainly in research, development and teaching, so as to fully employ everyone in his or her capacity. No; the first step towards political radicalisation is to question the nature, the significance and the relevance *of science itself as it is practised now*, and to question thereby the role of scientific workers.

Scientific workers are both the products and the victims of the capitalist division of labour. They can cease being the victims only if they refuse to be its products, to perform the role they are given and to practise this kind of esoteric and compartmentalised science. How can they refuse this? As a matter of principle, by refusing to hold a professional monopoly of expertise and by struggling for the reconquest and reappropriation of science by the people. The few Western examples of a successful implementation of this line of action usually draw inspiration from the Vietnamese and Chinese experience. The most important aspect of this experience is the following moral and political option: the goal is not the highest possible professional standards of a few specialists but, instead, the general progress and diffusion of knowledge within the community and the working class as a whole. *Any progress in knowledge, technology and power that produces a lasting divorce between the experts and the non-experts must be considered bad. Knowledge, like all the rest, is of value only if it can be shared.* Therefore, the best possible ways of sharing new knowledge must be the permanent concern of all research scientists. This concern will deeply transform the orientations of research and of science itself, as well as the methods and objects of scientific research. It will call for research to be carried out in constant co-operation and interchange between experts and non-experts.

These basic principles must be seen to be radical negations of the basic values of capitalist society. They imply that what is best is what is accessible to all. Our society, on the contrary, is based on the principle that what is best is whatever enables one individual to prevail over all others. Our whole culture — that is, science as well as patterns of consumption and behaviour — is based on the myth that everyone must prevail somehow over everyone else, and therefore that what is good enough for all is no good for anyone. A communist culture, on the contrary, is based on the principle that what is good for all of us is best for each and everyone of us.

There can be no classless society unless this principle is applied in all fields, including the field of science and knowledge. Conversely, science can cease to be bourgeois culture only if it is not only put at the service of the people, but becomes the people's own science, which means that science will be transformed in the process of its appropriation by the people. Indeed, as it is, science can never become the people's own science or science for the people; you cannot make a compartmentalised and professionalised elite culture into the people's own. Science for the people means the subversion of science as it is. As Hilary and Steven Rose put it:

> this transformation carries with it the breaking down of the barrier between expert and non-expert; socialist forms of work within the laboratory, making a genuine community instead of the existing degraded myth, must be matched by the opening of the laboratories to the community. The Chinese attempts to obliterate the distinction of expertise, to make every man his own scientist,* must remain the aim.[1]

---

*In 1976 we know better than to assume that 'man' equals humanity (eds).

# 5

# Contradictions of Science and Technology in the Productive Process

## *Mike Cooley*

Any meaningful analysis of scientific abuse must probe the very nature of the scientific process itself, and the objective role of science within the ideological framework of a given society. As such, it ceases to be merely a 'problem of science' and takes on a political dimension. It extends beyond the important, but limited, introverted soul-searching of the scientific community, and recognises the need for wider public involvement. Many 'progressive' scientists now realise that this is so, but still see their role as the interpreters of the mystical world of science for a largely ignorant mass, which when enlightened will then support the scientists in their intention 'not to use my scientific knowledge or status to promote practices which I consider dangerous'.

Those who, in addition to being 'progressive' have political acumen, know that a 'Lysistrata movement', even if it could be organised, is unlikely to terrify monopoly capitalism into applying science in a socially responsible manner. 'Socially responsible' science is only conceivable in a politically responsible society. That must mean changing the one in which we now live.

One of the prerequisites for such political change is the rejection of the present basis of our society by a substantial number of its members, and a conscious political force to articulate that contradiction as part of a critique of society as a whole. The inevitable misuse of science, and its consequent impact upon the lives of an ever-growing mass of people, provides the fertile ground for such a political development. It should constitute an important weapon in the political software of any conscious revolutionary.

Even Marxist scientists seem to reflect the internal political inces-
tuousness of the scientific community, and demonstrate in practice a
reluctance to raise these issues in the mass movement. Thus the debate
has tended to be confined to the rarified atmosphere of the campus,
the elitism of the learned body or the relative monastic quiet of the
laboratory.

Clearly, the ruling class, which has never harboured any illusions
about the ideological neutrality of science, will not be over-concerned
by this responsible disquiet. The Geneens of ITT and the Weinstocks
of GEC do not tremble at the pronouncements of Nobel Laureates.
It is true of course that the verbal overkill of the ecologist has rever-
berated through the quality press and caused some concern – not all of
it healthy – in liberal circles. But the working class, those who have it
within their power to transform society, those for whom such a transformation is an objective necessity, have not as yet been really involved.
Yet their day-to-day experience at the point of production brutally
demonstrates that a society which strives for profit maximisation is
incapable of providing a rational social framework for technology
(which they see as applied science).

'Socially irresponsible' science not only pollutes our rivers, air and
soil, provides CS gas for Northern Ireland, produces defoliants for
Vietnam and stroboscopic torture devices for police states. It also
degrades, both mentally and physically, those at the point of production, as the objectivisation of their labour reduces them to mere
machine appendages. The financial anaesthetic of the 'high-wage (a lie
in any case) high-productivity low-cost economy' has demonstrably
failed to numb workers' minds to the human costs of the fragmented
dehumanised tasks of the production line.

There are growing manifestations in the productive superstructure of
the irreconcilable contradictions at the economic base. The sabotage of
products on the robot-assisted line at General Motors Lordstown plant
in the United States, the 8 per cent absentee rate at Fiat in Italy, the
'quality' strike at Chryslers in Britain and the protected workshops in
Sweden reveal but the tip of a great international iceberg of seething
industrial discontent. That discontent, if properly handled, can be
elevated from its essentially defensive, negative stance into a positive
political challenge to the system as a whole.

The objective circumstances for such a challenge are developing
rapidly as the crushing reality is hammered home by the concrete
experience of more and more workers in high capital, technologically
based, automated or computerised plants. In consequence, there is a

gradual realisation by both manual and staff workers that the more elaborate and scientific the equipment they design and build, the more they themselves become subordinated to it, that is to the objects of their own labour. This process can only be understood when seen in the historical and economic context of technological change as a whole.

## SCIENCE AND THE CHANGING MODE OF PRODUCTION

The use of fixed capital, that is, machinery and, latterly, computers, in the productive process marked a fundamental change in the mode of production. It cannot be viewed merely as an increase in the rate at which tools are used to act on raw material. The hand tool was entirely animated by the workers, and the rate at which the commodity was produced – and the quality of it – depended (apart from the raw materials, market forces and supervision) on the strength, tenacity, dexterity and ingenuity of the worker. With fixed capital, that is the machine, it is quite the contrary in that the method of work is transformed as regards its use value (material existence) into that form most suitable for fixed capital. The scientific knowledge which predetermines the speeds and feeds of the machine, and the sequential movements of its inanimate parts, the mathematics used in compiling the numerical control programme, do not exist in the consciousness of the operator; they are external and act through the machine as an alien force. Thus science, as it manifests itself to the workers through fixed capital, although it is merely the accumulation of the knowledge and skill now appropriated, confronts them as an alien and hostile force, and further subordinates them to the machine. The nature of their activity, the movements of their limbs, the rate and sequence of those movements – all these are determined in quite minute detail by the 'scientific' requirements of fixed capital. Thus objectivised labour in the form of fixed capital emerges in the productive process as a dominating force opposed to living labour. We shall see subsequently when we examine concrete situations at the point of production that fixed capital represents not only the appropriation of *living* labour, but in its sophisticated forms (computer hardware and software) appropriates the scientific and intellectual output of the white-collar workers whose own intellects oppose them also as an alien force.

The more therefore that workers put into the object of their labour, the less there remains of themselves. The welder at General Motors who takes a robotic welding device and guides its probes through the welding procedures of a car body is on the one hand building skill into the machine, and deskilling themselves on the other. The accumulation of years of welding experience is absorbed by the robot's self-program-ming systems and will never be forgotten. Similarly, mathematicians working as stressmen in an aircraft company may design a software package for the stress analysis of airframe structures and suffer the same consequences in their jobs. In each case they have given part of themselves to the machine and in doing so have conferred life on the object of their labour — but now this life no longer belongs to them but to the owner of the object.

Since the product of their labour does not belong to the workers, but to the owner of the means of production in whose service the work is done and for whose benefit the product of labour is produced, it necessarily follows that the object of the workers' labour confronts them as an alien and hostile force, since it is used in the interests of the owner of the means of production. Thus this 'loss of self' of the worker is but a manifestation of the fundamental contradictions at the eco-nomic base of our society. It is a reflection of the antagonistic contra-diction between the interest of capital and labour, between the exploiter and the exploited. Fixed capital, therefore, at this historical stage, is the embodiment of a contradiction, namely that the means which could make possible the liberation of the workers from routine, soul-destroying, back-breaking tasks, is simultaneously the means of their own enslavement.

It is therefore obvious that the major contradiction can only be resolved when a change in the ownership of the means of production takes place. Much less obvious, however, is whether there exists a contradiction (non-antagonistic) between science and technology in their present form and the very essence of humanity. It is quite con-ceivable that our scientific methodology, and in particular our design methodology, has been distorted by the social forces that give rise to its development. The question therefore must arise whether the pro-blems of scientific development and technological change, which are *primarily* due to the nature of our class-divided society, can be solved solely by changing the economic base of that society.

The question is not of merely theoretical interest. It must be a burn-ing issue in the minds of those in Vietnam who are responsible for their country's programme of reconstruction. It must be of political concern to those in China, to establish if Western technology can be simply applied to a socialist society. Technology at this historical stage, in a class-divided society, such as Britain, is the embodiment of two oppo-

sites — the possibility of freeing the workers, yet the actuality of ensnar-
ing them. The possibility can only become actuality when the workers
own the object of their labour. Because the nature of this contradiction
has not been understood, there have been the traditional polarised
views: 'technology is good'; and 'technology is bad'. These polarised
views are of long standing and not merely products of space-age technology. From the earliest times a view has persisted that the introduction of mechanisation and automated processes would automatically
free people to engage in creative work. This view has persisted as
consistently in the field of intellectual work as it has in that of
manual labour. As far back as 1624, when Pascal introduced his first
mechanical calculating device he said, 'I submit to the public a small
machine of my own invention, by means of which you alone may
without any effort perform all the operations of arithmetic and may be
relieved of the work which has so often fatigued your spirit when you
have worked with the counters and with the pen.' Only twenty-eight
years earlier in 1596 an opposite view was dramatically demonstrated
when the city council of Danzig hired an assassin to strangle the inventor of a labour-saving ribbon-loom, a defensive if understandable
attempt, repeated time and again in various guises during the ensuing
500 years to resolve a contradiction at an industrial level when only a
revolutionary political one would suffice. It is of course true that the
contradiction manifests itself in industrial forms even to this day.

## THE OBSOLESCENCE OF FIXED CAPITAL

There is first the ever shorter life of fixed capital (the increasing rate of
obsolescence of machinery). Early wheeled transport existed in that
form for thousands of years; steam-engines made by Boulton and Watt
two hundred years ago were still operating about a hundred years later;
a century ago, when an employer purchased a piece of machinery, he
could rest assured that it would last his lifetime and would be an asset
he could pass on to his son.

In the 1930s machinery was obsolete in about twenty-five years,
during the 1950s in ten years, and at the moment computerised equipment is obsolete in about three to five years. Then there is the growing
volume of fixed capital necessary to provide the total productive environment for a given commodity — the cost of the total means of production is ever-increasing. That is not to say that the cost of individual
commodities will continue to increase. The most complicated lathe one

could get 100 years ago would have cost the equivalent of ten workers' wages per annum. Today, a lathe of comparable complexity, with its computer-tape control and the total environment necessary for the preparation of those tapes and the operation of the machine, will cost something in the order of a hundred workers' wages per annum. The industrial manifestations of the contradiction now begin to emerge very clearly indeed. Confronted with equipment which is getting obsolete literally by the minute, and has involved enormous capital investment, the employer will seek to recoup his investment by exploiting that equipment for twenty-four hours per day. In consequence of this, employers will seek to eliminate all so-called non-productive time, such as tea breaks, will seek to subordinate the employees more and more to the machine in order to get the maximum performance, and will insist that the equipment is either worked upon on three shifts to attain a twenty-four hour exploitation, or is used on a continuous over-time basis. This trend has long since been evident in the manual field on the workshop floor. It is now beginning to be a discernible pattern in a whole range of white-collar occupations.

## THE PROLETARIANISATION OF INTELLECTUAL WORKERS

An analysis of this problem in British companies demonstrates that employers will wish to ensure that all their white-collar employees who use this kind of equipment accept the same kind of subordination to the machine that they have already established for manual workers on the shop floor. To say that this is so is not to make a prediction about the far-distant future. In 1971 my union (AUEW – TASS) was involved in a major dispute with Rolls Royce, which cost the union £250,000. The company sought, amongst other things, to impose on the design staff at the Bristol plant the following conditions:

> The acceptance of shift work in order to exploit high capital equipment, the acceptance of work measurement techniques, the division of work into basic elements, and the setting of times for these elements, such time to be compared with actual performance.

In this instance the union was able, by industrial action, to prevent the company from imposing these conditions. They are, however, the sort of conditions which employers will seek increasingly to impose upon the white-collar workers. When staff workers, whether they be technical, administrative or clerical, work in a highly synchronised, computerised environment, the employer will seek to ensure that each element of their work is ready to feed into the process at the precise time at which it is required. Mathematicians, for example, will find that they have to have their work ready in the same way as a Ford worker has to have the wheel ready for the car as it passes him on the production line. In consequence of this, many graduates, who in the past would never have recognised the need to belong to a real trade union, now find that they need the same kind of bargaining strength that manual workers have accepted on the shop floor for some considerable length of time. *In fact, one can generalise and say that the more technological change and computerisation enters white-collar areas, the more workers in those areas will become proletarianised.* The consequence of this will not be limited to the work situation. They will spread right across the family, social and cultural life of the white-collar worker. Consider the consequences of shift working for example. In a survey carried out in West Germany it was demonstrated that the ulcer rate amongst those working a rotating shift was eight times higher than amongst other workers. Other surveys have shown that the divorce rate amongst shift workers is approximately 50 per cent higher than normal, whilst the juvenile delinquency rate of their children can often be 80 per cent higher. There are a whole series of examples in Britain of the manner in which the cultural and social life of AUEW – TASS members has been disrupted by the introduction of this kind of equipment.

Thus, whilst it is true that automated and computerised equipment *could* free people from routine, soul-destroying, back-breaking tasks, and free them to engage in more creative work, the reality in our profit-orientated society is that in many instances it actually lowers 'the quality of life'.

There are also good grounds for assuming that automated and computerised systems will in many instances diminish rather than enhance the creativity of scientific and technological workers. Computer Aided Design (CAD) is a useful occupational aperture through which to view a scenario that will become commonplace to many in the next few years.

In selling the idea of computers to the design community, it is suggested that the computer will merely deal with the quantitative factors and the designer will deal with value judgements and the creative elements of the design process. It is of course true that the design process is, amongst other things, an interaction of the quantitative and the qualitative. It is not, however, true that design methodology is such that these can be separated into two disconnected elements which can then be applied almost as chemical compound. The process by which these two opposites are united by the designer to produce a new whole is a complex and as yet ill-defined and ill-researched area.

The sequential basis on which the elements interact is of extreme importance. The nature of that sequential interaction, and indeed the ratio of the quantitative to the qualitative, depends on the commodity under design consideration. Even where an attempt is made to define the proportion of the work that is creative, and the proportion that is non-creative, what cannot be readily stated is the stage at which the creative element has to be introduced when a certain stage of the non-creative work has been completed. The very subtle process by which designers review the quantitative information they have assembled, and then make the qualitative judgement, is extremely complex. Those who seek to introduce computerised equipment into this inter-action attempt to suggest that the quantitative and the qualitative can arbitrarily be divided, so that the computer handles the quantitative. (This is in reality a devious introduction of 'Taylorism' into advanced technological work — an attempt to further subdivide an 'intellectual activity' into its 'manual' and 'intellectual' components.)

Since CAD dramatically increases the rate at which the quantitative is handled, a serious distortion of this dialectical interaction takes place, frequently to the detriment of the qualitative. There are there-fore good grounds for assuming that the crude introduction of the computer into the design process, in keeping with the Western ethic of 'the faster the better', may well result in a deterioration of the design quality. It is typical of the narrow, fragmented and short-term view which capitalism takes of all productive processes, that these important philosophical considerations are ignored. Much design research is limited to considerations of design techniques and associated hardware and software with precious little regard for the objective requirements of the design staff or, more importantly, the public. Such research accurately reflects our economic base: equipment and hence capital first; people last.

Elitist designers, steeped in their traditional professionalism, believed (and many still believe!) that their creative talents provide an eternal occupational sanctuary against the creeping proletarianisation of all white-collar workers. Architects, for example, conceded that there might be problems for aircraft designers, mechanical designers or civil engineers, but not for them. After all, is not architecture the 'Queen of the Arts rather than the Father of Technology'? However, capitalism, in its relentless drive to exploit all who work, has not forgotten the architects. For them there has been specifically produced a software package known (appropriately) as HARNESS. The concept behind this system is that the design of buildings can be systematised to such an extent that each building is regarded as a communication route. Stored within the computer system are a number of predetermined architectural elements which can be disposed around the communication route on a Visual Display Unit to produce different building configurations. Only these predetermined elements may be used and architects are reduced to operating a sophisticated 'Lego' set. Their creativity is limited to choosing how the elements will be disposed rather than considering in a panoramic way the types and forms of elements which might be used. As Marx pointed out in *Capital*: 'a bee puts to shame many an architect in the construction of her cells. But what distinguishes the worst architect from the best of bees is this, the architect raises his structure in imagination before he erects it in reality. At the end of every labour process, we get a result that already existed in the imagination of the labourer at its commencement'.[1]

It is clear that HARNESS will reduce the distinction between the architect and the bee and that capitalism will insist that in future architects will work in a more 'bee-like' fashion! This will gradually apply to all technological and scientific workers, whatever systems are devised to control their 'mode of intellectual production' in the same way as manual workers on the shop floor are controlled. Employers have long sought ways and means of controlling 'their' elusive, creative technical and white-collar staff. The computerised system is one Trojan Horse widely used to do so. The process is succinctly described in the magazine *Realtime* by a writer fresh from an IBM customer training ing course:

Now an operating system is a piece of software, functionally de-signed to do most efficiently a particular job — or is it? It gradually dawned on me that some rather obnoxious cultural assumptions have been imported lock, stock and barrel into IBM software. Insidious, persuasive assumptions, which appear to be a natural product of logic — but are they? The whole thing is a complete totalitarian hierarchy. The operating system runs the computer installation. The chief and most privileged element is the 'Supervisor'. Always resident in the most senior position in main storage, it controls,

through its minions, the entire operation. Subservient to the supervisor is the bureaucratic machinery — job management routines, task
management, input/output scheduling, space management and so on.
The whole thing is thought out as a rigidly controlled, centralised
hierarchy. And as machines get bigger and more powerful, so the
operating system grows and takes more powers. One lecturer soared
into eloquence in comparing the various parts of the operating
system to the directors, top management, middle management, shop
foremen and ordinary pleb workers of a typical commercial company. In fact, the whole of IBM terminology is riddled with class
expressions — master files, slave cylinders, high and low level languages, controller, scheduler, monitor.[2]

The same writer then generalised some of the contradictions of centralised operating systems. (These coincided closely with my own findings
when I investigated the contradiction in the specific field of Computer
Aided Design.)

The drawbacks of the centralised operating system are many. It is a
constraining and conservative force. A set of possibilities for the
computer system is chosen at a point in time, and a change involves
regeneration of the system. It imposes conformity on programming
methods and thought. Another amazingly apt quote from an IBM
lecturer — 'Always stick to what the system provides, otherwise you
may get into trouble'. It mystifies the computer system by putting
its most vital functions into a software package which is beyond the
control and comprehension of the applications programmer, thus
introducing even into the exclusive province of Data Processing the
division between software experts and other programmers, and
re-inforcing the idea that we do not really control the tools we use,
but can only do something 'if the operating system lets you' — a
phrase which I am sure many of us have used. The system which
results seems absurdly top heavy and complex. The need to have
everything centrally controlled seems to impose an enormous strain.

The introduction of computerised systems is frequently used as a
smokescreen to introduce another management control weapon — job
evaluation. Pseudo-scientific reasons are given for fragmenting jobs and
slotting the subdivided function into a low level of the system hierarchy
with correspondingly low wages for 'appropriate' job grades. My experience of this in industry tends to show that it is frequently actually
used to consolidate the unequal pay and opportunities of women by

either implying (they can no longer state it openly as they have done in the past), or ensuring by structural means and recruitment, that the fragmented functions are 'women's work'. There is of course no such thing as 'women's work' in this sense any more than there is women's mathematics, women's physics, women's literature or women's music. There is only work — the means by which employers extract profits from all of us but higher profits from women. Thus a contradiction will exist in that although scientific and technological progress could provide the objective circumstances for greater equality between the sexes in the productive process, in our profit-orientated society the reverse will frequently be the case. Women will have to fight not only the traditional forms of discrimination, but much more sophisticated and scientifically structured ones with little indication that the unions catering for such workers have really understood the nature or scale of these problems.

The emergence of fixed capital as a dominant feature in the productive process means that the organic composition of capital is increased and industry becomes capital intensive rather than labour intensive. Human beings are increasingly replaced by machines. This in itself increases the instability of capitalism; on the one hand, capitalism uses the quantity of working time as the sole determining element, yet at the same time continuously reduces the amount of direct labour involved in the production of commodities. At an industrial level, literally millions of workers lose their jobs and millions more suffer the nagging insecurity of the threat of redundancy. An important new political element in this is the class composition of those being made redundant. Just as the use of high-capital equipment has spread out into white-collar and professional fields so also the consequences of high-capital equipment do likewise. Scientists, technologists, professional workers, clerical workers, all now experience unemployment in a manner that only manual workers did in the past. Verbal niceties are used to disguise their common plight. A large West London engineering organisation declared its scientists and technologists 'technologically displaced', its clerical and administrative workers 'surplus to requirements', and its manual workers 'redundant'. In fact they had all got the good old-fashioned sack! In spite of different social, cultural and educational backgrounds, they all had a common interest in fighting the closure of that plant, and they did. Scientists and technologists paraded around the factories carrying banners demanding 'the right to work' in a struggle that would have been inconceivable a mere ten years ago. Technological change was indeed proletarianising them. In consequence of the massive and synchronised scale of production which modern technology requires, redundancies can affect whole communities. During a recession in the American aircraft industry in the early 1970s a union banner read: 'Last out of Seattle please put the lights out.' Because of this change in the organic composition of capital, society is gradually being conditioned to accept the idea of a permanent pool of unemployed people. Thus we find in the United States, in spite of the artificial stimulus of the Vietnam War, over the past ten years about five million people have been permanently out of work.

We have witnessed in this country the large-scale unemployment of recent years. Unemployment is considerable in Italy, and even in the West German miracle there are sections of workers — particularly over the age of fifty — who are now experiencing long terms of unemployment. This unemployment itself creates contradictions for the ruling class. It does so because people have a dual role in society, that of producers and consumers. When you deny them the right to produce, you also limit their consumption power. In an attempt to achieve a balance, efforts are now being made to restructure the social services to maintain that balance between unemployment and the purchasing power of the community. In the United States, President Kennedy spoke of a 'tolerable level of unemployment'. In Britain in the 1960s Harold Wilson, stoking the fires of industry with the taxpayer's money through the Industrial Reorganisation Corporation to create the 'white heat of technological change' spoke in a typical double negative of a 'not unacceptable level of unemployment'.

A remarkable statement for a so-called socialist Prime Minister! The net result is that there is on the one hand an increased work tempo for those in industry, whilst on the other hand there is a growing dole queue, with all the degradation that that implies; nor is there any indication that the actual working week has been reduced during this period. Indeed, in spite of all the technological change since the war, the actual working week in Britain for those who have jobs is longer now than it was in 1946. Yet the relentless drive goes on to design machines and equipment which will replace workers. Those involved in such work seldom question the nature of the process in which they are engaged. Why, for example, the frantic efforts to design robots with pattern recognition intelligence when we have a million and a half people in the dole queue in Britain whose pattern recognition intelligence is vastly greater than anything yet conceived even at a theoretical level?

The system seeks in every way to break down the workers' resistance to being sacked. One of the sophisticated devices was the Redundancy Payments Act under the 1964–70 Labour Government. Practical experience of trade unions in Britain demonstrates that the lump sums involved broke up the solidarity at a number of plants where struggle was taking place against a closure.

A much more insidious device is to condition workers into believing that it is their own fault that they are out of work, and that they are in fact unemployable. This technique is already widespread in the United States, where it is asserted that certain workers do not have the intelligence and the training to be employed in modern technological society. This argument is particularly used against coloured workers, Puerto Ricans and poor whites. There is perhaps here fertile ground for some of the 'objective research' of Jensen and Eysenck.

The concept of a permanent pool of unemployed persons as a result of technological change also brings with it the danger that those unemployed would be used as a disciplining force against those still in work. It undoubtedly provides a useful pool from which the army and police force can draw, and during the recent redundancies in Britain, a considerable number of redundant workers from the North-east were recruited into the army and then used against workers in Northern Ireland. Coupled with the introduction of this high-capital equipment is usually a restructuring known as 'rationalisation'. The epitome of this in Britain is the GEC complex with Arnold Weinstock at its head. In 1968, this organisation employed 260,000 workers and made a profit of £75 million. In consequence of quite brutal redundancies, the company's work force was reduced to 200,000, yet profits went up to £105 million. These are the kind of people who are introducing high-capital equipment, and they make their attitude to human beings absolutely clear. It is certainly profits first and people last! One quotes Arnold Weinstock not because he is particularly hideous (he is in fact extremely honest, direct and frank) but because he is prepared to say what others think. He said on one occasion that 'People are like elastic, the more work you give them, the more they stretch.' We know, however, that when people are stretched beyond a limit, they break. My union has identified a department in a West London engineering company where the design staff were reduced from thirty-five to seventeen and there were six nervous breakdowns in eighteen months. Yet people like Weinstock are held up as a glowing example to all aspiring managers. One of his own senior managers once boasted that 'he takes people and squeezes them till the pips squeak'. I think it is a pretty sick and decaying society that will boast of this kind of behaviour.

Most industrial processes, however capital intensive they might be, still require human beings in the total system. Since highly mechanised or automated plant frequently is capable of operating at very high speeds, employers view the comparative slowness of human beings in their interaction with the machinery as a bottleneck in the over-all system. In consequence of this, pay structures and productivity deals are arranged to ensure that the workers operate at an ever faster tempo.

In many instances the work tempo is literally frantic. In one automobile factory in the Midlands in Britain they reckon that they 'burn a man up' on the main production line in ten years. They recently tried to get our union to agree that nobody would be recruited for this type of work over the age of thirty. For the employer it is like having a horse or dog. If you must have one at all, then you have a young one so that it is energetic and frisky enough to do your bidding all the time. So totally does the employer seek to subordinate the worker to production, that he asserts that the worker's every minute and every movement 'belong' to the employer. Indeed, so insatiable is the thirst of capital for surplus value, that it thinks no longer in terms of minutes of workers' time, but fractions of minutes. The grotesque precision with which this is done to workers can be seen from a report which appeared in the *Daily Mirror* of 7 June 1973. It gave particulars of the elements which make up the 32.4 minute rest-allowance deal for body press workers on the Allegro car: trips to the lavatory 1.62 minutes (note; not 1.6, not 1.7, but 1.62!): recovery from fatigue 1.30 minutes, sitting down after standing too long sixty-five seconds, for monotony thirty-two seconds. The report went on to point out that, in a recent dispute, the workers sought to increase the monotony allowance by another sixty-five seconds! The methods may vary from company to company, or from country to country, but where the profit motive reigns supreme, the degradation and subordination of the worker is the same. George Friedmann has written of two different methods used by great French companies, Berliot in Lyons and Citroen in Paris:

> Why has the Berliot works the reputation, in spite of the spacious beauty of its halls, of being a jail? Because here they apply a simplified version of the Taylor method of rationalising labour, in which the time taken by a demonstrator, an 'ace' worker, serves as the criterion imposed on the mass of workers. He it is who fixes, watch in hand, the 'normal' production expected from a worker. He seems when he is with each worker, to be adding up in an honest way the time needed for the processing of each item. In fact if the worker's movement seem to him to be not quick or precise enough, he gives a practical demonstration, and his performance determines the norm expected in return for the basic wage. Add to this supervision in the technical sphere the disciplinary supervision of uniformed warders to patrol the factory all the time and go as far as to push open doors of the toilets to check that the men squatting there are not smoking, even in workshops where the risk of fire is non-existent.
>
> At Citroen's the methods used are more subtle. The working teams are in rivalry with one another, the lads quarrel over travelling cranes, drills, pneumatic grinders, small tools. But the supervisors in white coats, whose task is to keep up the pace, are insistent, pressing, hearty. You would think that by saving time a worker was doing them a personal favour. But they are there, unremittingly on the back of the foreman, who in turn is on your back; they expect you to show an unheard of quickness in your movements, as in a speeded-up motion picture! Within this context, the desire of companies to recruit only those under the age of 30 can be seen in its dehumanised context.[3]

Although this is the position on the workshop floor, it would be naive indeed to believe that the use of high-capital equipment will be any more liberating in the fields of clerical, administrative, technical, scientific and intellectual work.

Age limits are now gradually being introduced in the white-collar areas. In 1971 the *Sunday Times* gave a list of the peak-performance ages for mathematicians, engineers, physicists and others. For some of these the peak-performance age was twenty-nine and thirty. It has been suggested that in order to utilise this high-capital equipment as effectively as possible, a careers profile should be worked out for those who have to interface with it.

When workers reach their peak-performance age, it is suggested that this should be followed by a careers plateau for three or four years and thereafter, unless the employee has moved into management, that they be subjected to a 'careers de-escalation'. The obvious extension of the careers de-escalation is redundancy. Practical experience demonstrates, particularly during periods of redundancy, that older people are being eliminated in this way. They are being eliminated or down-graded to lower-paid work simply because they have committed the hideous crime of beginning to grow old. We are, as Samuel Beckett once said, 'all born of the gravedigger's forceps'. Growing old is the most natural human process. It is a biological process, but, in the contradictory nature of our profit-orientated society, it is treated almost as a crime. It is true that the kind of equipment we have been discussing imposes very stringent demands upon those who have to interface with it. Seen in terms of the total man/machine systems, people are slow, inconsistent, unreliable, but still highly creative. The machine is the dialectical opposite, in that it is fast, reliable, consistent, but totally non-creative. As people attempt to respond to the machine, enormous stress is placed upon them. My union has identified areas within the design activity where by using interactive graphic systems the decision-making rate of the designer is increased by 1900 per cent.

## HUMANS AS MACHINES

Again there are analogies to be drawn from the shop floor. In the British Steel Corporation a productivity agreement has introduced medical checks. In practice these medical checks meant the operators were tested to ensure that their response rates were fast enough to interface with the equipment. They were merely tested for their response rates as a diode might be. A series of occupational suitability tests and character compatibility assessments are now gradually being used to do the same sort of thing to white-collar workers who have to use high-capital equipment. The object is to transform the worker, whether by hand or brain, into a suitable machine appendage. To do this, all the human requirements of the individual must be denied. They must be transformed into operating units. The 'scientific' manner in which this man/machine interface is planned emphasises the total dehumanisation of the so-called technologically advanced production techniques. Robert Boguslaw has recently pointed out how some behavioural scientists view the human being in this situation:

Our immediate concern let us remember, is the explication of the operating unit approach to system design, no matter *what* materials are used. We must take care to prevent this discussion from degenerating into a single-sided analysis of the complex characteristics of one type of system material: namely human beings. What we need is an inventory of the way in which human behaviour can be controlled and a description of some instruments which will help us to achieve control. If this provides us with sufficient 'handles' on human materials so that we can think of them as one thinks of metal parts, electrical power, or chemical reactions, then we have succeeded in placing human material on the same footing as any other materials and can proceed with our problems of systems design.[4]

This, then, is the objective dehumanisation which takes place side by side with this advanced technology.

There are however, many disadvantages in the use of human operating units. They are somewhat fragile; they are subject to fatigue, obsolescence, disease and death; they are frequently stupid, unreliable and limited in memory capacity. But beyond all this, they sometimes seek to design their own circuitry. This, in a material, is unforgiveable. Any system utilising them must devise appropriate safeguards.

Thus, if workers use their greatest attribute — that is, their ability to think — their ability to design their own circuitry — this is regarded as disruptive. The objective requirement of industry, then, is for people who will act as robots, people who are interchangeable with robots. Some scientists and technologists take the smug view that this can only happen in any case to mere manual workers on the shop floor. They fail to realise that the problem is now at their own doorstep. At a conference on robot technology at Nottingham University in April 1973, a programmable draughting or design system was accepted by definition as being a robot. One of the manufacturers of robotic equipment pointed out that 'Robots represent industry's logical search for an obedient workforce.' This is a very dangerous philosophy indeed. The great thing about people is that they are sometimes disobedient. Most human development, technical, cultural and political, depended upon those movements which questioned, challenged and where necessary disobeyed the then established order.

The ruling class views all workers, whether by hand or brain, as units of production. Only when that class reality has been firmly grasped can the chasm which divides the potentialities of science and technology from the current reality be understood. The gap between that which is possible, and that which is, widens daily. Technology can produce a Concorde but not enough simple heaters to save the hundreds of old-age pensioners who each winter die in London of hypothermia. Only when one realises that the system regards old-age pensioners as discarded units of production does this make sense — capitalist sense. This is part of their social design, and from a ruling-class viewpoint is quite 'scientific' and abides closely by the principles observed in machine design. I know, as a designer, that when you design a unit of production you ensure that you design it to operate in the minimum environment necessary for it to do its job. You seek to ensure that it does not require any special temperature-controlled room unless it is absolutely essential. In designing the lubrication system you do not specify any exotic oils as lubricants unless it is necessary. You ensure that its control system is provided with the minimum brain necessary for it to do its job. You do not, for example, have a machine tape con-trolled if you can get away with a manual one. Finally, you provide it with the minimum amount of maintenance; in other words, you design for it the maximum life span in which it will operate before a failure.

Those who control our society see human beings in the same way. The minimum environment for people means that you provide them with the absolutely lowest level of housing which will keep them in a healthy enough state to do their job. If one doubts that, it is still worth remembering that 7,000,000 people live in slums in Britain. The equiva-lent of fuel and lubrication for the machine is the food provided for a person. This is also kept at a minimum for those who work. We even find Oxford dieticians still telling old-age pensioners how they can manage on £2 of food per week. The minimum brain is provided by an educa-tional system which gives people enough knowledge to do their job, which trains them to do their job, but does not educate them to think about their predicament or that of society as a whole.

The minimum maintenance necessary is provided through the National Health Service, which concentrates on curative rather than preventive medicine, and the reality, the harsh reality, is that when people have finished their working life, they are thrown on the scrap heap like an obsolete machine.

If that is felt to be an extreme position, it is worth recalling the statement of the doctor at Willesden Hospital, who said there was no need for National Health patients over the age of sixty-five to be resuscitated (the doctor was actually sixty-eight!). When a barrage of protest was raised the statement was hurriedly withdrawn as a mistake! The real mistake he made was to reveal in naked print one of the under-lying assumptions of our class-divided society. Science and technology cannot be humanely applied in an inherently inhuman society, and the contradictions for scientific workers in the application of their abilities will grow and, if properly articulated, will lead to a radicalisation of the scientific community.

A source of great stress, particularly for white-collar workers, is the problem of knowledge obsolescence. This problem is closely related to the rate at which technology itself is changing. It seems desirable to attempt to quantify technological change.

The scale of technological development in the last twenty years is probably equal to that in all of humanity's previous existence. The scale of scientific effort, which is closely related to technological change, has in the present century increased out of all recognition. Bernal calculated that in 1896 there were perhaps in the whole world some 50,000 people who between them carried on the whole tradition of science, not more than 15,000 of whom were responsible for the advancement of knowledge through research. Today, the total number of scientific workers in industry, government and academic circles in Britain alone is over 400,000. This is merely a reflection in manpower of the statistics of the actual rate of technological change, which in the last century alone has meant that our speed of communication was increased by $10^7$, our speed of travel by $10^2$, data-handling by $10^6$, energy resources by $10^3$ and weapons power by $10^6$.

As the rate of technological change increases, so also does the rate at which knowledge becomes obsolete. Mathematical models described by Sir Frederick Warner indicate that in order to keep abreast of this knowledge engineers would have to spend 15 per cent of their time in up-dating his current knowledge. Mr Norman McRae, Deputy Editor of *The Economist*, stated in the January 1972 issue that

> The speed of technological advance has been so tremendous during the past decade that the useful life of knowledge of many of those trained to use computers has been about three years. [and, further] A man who is successful enough to reach a fairly busy job at the age of 30, so busy that he cannot take sabbatical periods for study, is likely at the age of 60 to have about one-eighth of the scientific (including business scientific) knowledge that he ought to have for the proper functioning in his job.

It has even been suggested that if one divided knowledge into quartiles of out-datedness, those in the age bracket over forty-five would find themselves in the same quartile as Pythagoras and Archimedes.

The stress that this places upon staff workers, in particular older people, should not be under-estimated. It is the responsiblity of the trade unions to protect these older people. This they should do not in any patronising, benevolent fashion, but in recognition of the class right of these older people to work at a civilised tempo. For these are the ones who in the past have created the real wealth that has made the purchase of this kind of high-capital equipment possible. All younger technologists should fully understand that however energetic and forceful they may feel now, they will inevitably begin to grow old, and if they allow older members to be treated in this way they are creating a framework of oppression which will be used against them in the future.

## THE FRAGMENTATION OF SKILLS

A major part of the process of technological change is the fragmentation of jobs into deskilled, narrow elements. It is also part of the historical division between intellectual and manual work. In the past, many jobs which were essentially manual did contain within them major elements of intellectual and scientific work. Sir William Fairbairn's definition of a millwright in 1861 illustrates the point.

> The millwright of former days was to a great extent the sole representative of mechanical art... he was an itinerant engineer and mechanic of high reputation. He could handle the axe, the hammer and the plane with equal skill and precision; he could turn, bore or forge with the despatch of one brought up to these trades and he could set out and cut furrows of a millstone with an accuracy equal or superior to that of the miller himself .... Generally he was a fair mathematician, knew something of geometry, levelling and mensuration, and in some cases possessed a very competent knowledge of practical mathematics. He could calculate the velocities, strength and power of machines, could draw in plan and section, and could construct buildings, conduits, or water course in all forms and under all conditions required in his professional practice. He could build bridges, cut canals and perform a variety of work now done by civil engineers.[5]

All the intellectual work has been long since withdrawn from the mill-wright's function.

This fragmentation of skills now applies equally in the white-collar areas. The draughtsman of the 1930s in Britain was the centre of design. He could design the component, stress it, specify the materials to be used, define the method of lubrication, and write the test specifications. With the increasing complexity of technology, each of these have now been fragmented into narrow, specialised areas. The draughtsman draws, the stressmen carry out the calculations, the metallurgist specifies the materials, the tribologist decides upon the means of lubrication.

It has been common for some time to talk about 'dedicated machines'. It is now a fact that when defining a job function employers define a dedicated appendage to the machine, the operator.

Even our educational system is being distorted to produce these 'dedicated people for dedicated machines'. It is no longer a matter that the people are being educated to think; they are being trained to do a narrow, specific job. Much of the unrest amongst students is recognition that they are being trained as industrial fodder for the large monopolies in order to fit them into narrow fragmented functions where they will be unable to see in an over-all panoramic fashion the work on which they are engaged.

In order to ensure that the right kind of 'dedicated product' is turned out of the university, we find the monopolies attempting to determine the nature of university curricula and research programmes. Warwick University was a classical example. In particular, at research level, the monopolies increasingly attempt to determine the nature of research through grants which they provide to universities or research projects undertaken in their own laboratories. Many research scientists still harbour illusions that they are in practice 'independent, dedicated searchers of truth'.

The 'truth' for them has to coincide with the interests of the monopolies if they are to retain their jobs. William H. Whyte Jr pointed out that in the United States, out of 600,000 persons then engaged in scientific research, not more than 5000 were allowed to choose their research subject and less than 4 per cent of the total expenditure was devoted to 'creative research', which does not offer immediate prospects of profits.

He recognises the long-term consequences of this and concludes: 'If corporations continue to mould scientists the way they are now doing, it is entirely possible that in the long run this huge apparatus may actually slow down the rate of basic discovery it feeds on.'[6]

## PERSPECTIVES FOR REVOLUTIONARY ACTION

I have up to now concentrated on the contradictions as they affect the worker by hand or brain. There are of course problems for the employer, and an understanding of some of these is of considerable tactical importance.

One of the contradictions for the employer is that the more capital he accumulates in any one place, the more vulnerable it becomes. The more closely he synchronises his industry and production by using computers, the greater becomes the strike power of those employed in it. Mao Tse Tung once said, in his military writings, that the more capitalised an army becomes, the more vulnerable it becomes also. This has been demonstrated in Vietnam, where NLF cadres with £1.50 shells were able to destroy American aircraft with airborne computers costing something like £2.5 million each. A Palestinian guerilla with a revolver costing perhaps £20 can hijack a plane costing several million dollars and destroy it at some safe airfield. High-capital equipment, although it appears all powerful and invincible, always has a point of vulnerability, and possibilities for sabotage and guerilla warfare are considerable. A quite small force can destroy or immobilise plant equipment or weapons costing literally millions. The capitalisation of industry also produces an analogous situation. In the past, when a clerical worker went on strike, it had precious little effect. Now if the wages of a factory are carried out by a computer, a strike by clerical workers can disrupt the whole of the plant. It is also true on the factory floor that in the highly synchronised motor-car industry, a strike of twelve workers in the foundry can stop large sections of the entire motor-car industry.

The same is happening in the design area. As high-capital equipment, through Computer Aided Design, is being made available to design staffs, first it proletarianises them, but second it also increases their strike power. In the past, when draughtsmen went on strike, they simply put down their 6H pencils and their rubbers, and there was unfortunately a considerable length of time before an effect was felt upon production, even when the manual workers were blacking their drawings. With the new kind of equipment described, where computer tapes are being prepared or where high-capital equipment is used for interactive work, the effects of a strike will in many instances be immediate, and production will be affected in a very short length of time.

This will apply equally to hosts of other jobs and occupations, in banking, insurance, power generation, civil transport, as well as those more closely connected with industry and production.

Thus, whilst the introduction of fixed capital enables the employer to displace some workers and subordinate others to the machine, it also embodies within it an opposite in that it provides the worker with a powerful industrial weapon to use against the employer who introduced it.

This is even the case when industrial action short of strike action is taken. As has been pointed out, the activity of the worker is transformed to suit the requirement of fixed capital. The more complete that transformation, the greater is the disruptive effect of the slightest deviation by the worker from a predetermined work sequence. Industrial militants with an imaginative and creative view of industrial harassment have been able to exploit this contradiction by developing such techniques: 'working to rule', 'working without enthusiasm' and 'days of non-co-operation'. I know from personal experience that these techniques can reduce the output of both manual and staff workers by up to 70 per cent without placing on the workers involved the economic hardship of a full strike.

Since much of the sophisticated equipment I have described earlier is very sensitive and delicate in a scientific sense, it has to be handled with great care and is accommodated in purpose-built structures in conditions of clinical cleanliness. In many industries the care the employer will lavish on 'his' fixed capital is in glaring contrast with the comparatively primitive conditions of 'his' living capital. The campaign for parity with equipment, which perhaps started facetiously in 1964 with that placard at Berkeley which parodied the IBM punchcard ('I am a human being: Please do not fold, spindle or mutilate') has now assumed significant industrial dimensions. In June 1973 designers and draughtsmen members of the AUEW-TASS employed by a large Birmingham engineering firm, officially claimed 'Parity of environment with the CAD Equipment' in the following terms:

> This claim is made in furtherance of a long standing complaint concerning the heating and ventilation in the Design and Drawing Office Area going back to April 1972. Indeed to our certain knowledge these working conditions have been unsatisfactory as far back as 1958. We believe that if electromechanical equipment can be

considered to the point of giving it an air conditioned environment for its efficient working the human beings who may be interfaced with this equipment should receive the same consideration.

It is an interesting reflection on the values of advanced technological society that it subsequently took three industrial stoppages to achieve for the designers conditions approaching those of the CAD equipment. The exercise also helped to dispel some illusions about highly qualified design staff not needing trade unions.

Scientists must now begin to learn the lessons of such experiences, and to understand that their destiny is bound up with all of those 'moulded' by the systems. Only when they are prepared to be involved in political struggle with them, can they ever begin to move towards a society where scientists will be able to give 'according to their ability'. It is the historical task of the working class to effect such a transformation, but in that process scientists and technologists can be powerful and vital allies for the working class as a whole. This will mean that scientists will have to involve themselves in the political movement. Above all, they must attempt to understand that the products of their ingenuity and scientific ability will become the objects of their own oppression and that of the mass of the people until they are courageous enough to help form that sort of society

> When the enslaving subordination of the individual to the division of labour, and with it the antithesis between mental and physical work has vanished; when labour is no longer merely a means of life but has become life's principal need; when the productive forces have also increased with the all-round development of the individual and all the springs of co-operative wealth flow more abundantly. Only then will it be possible completely to transcend the narrow outlook of bourgeois right and only then will society be able to inscribe on its banners: From each according to his ability, to each according to his needs.[7]

Then, and then only will scientists be able to truly give of their ability to meet the needs of the community as a whole rather than maximise profits for the few.

# 6

# The Politics of Neurobiology: Biologism in the Service of the State

*Steven Rose and Hilary Rose*

Biologism is the attempt to locate the cause of the existing structure of human society, and of the relationships of individuals within it, in the biological character of the human animal. For biologism, all the richness of human experience and the varying historical forms of human relationships merely represent the product of underlying biological structures; human societies are governed by the same laws as ape societies, the way that an individual responds to his or her environment is determined by the innate properties of the DNA molecules to be found in brain or germ cells. In a word, the human condition is reduced to mere biology, which in its turn is no more than a special case of the laws of chemistry and hence of physics.

As a theoretical model, biologism is thus a form of that reductionism which is the dominant paradigm of contemporary Western science. As a philosophy, reductionism's premises are that

(*a*) Sciences are arranged in a hierarchical order, varying from high-level disciplines such as economics and sociology to lower-level ones, such as biology, chemistry and, at the base, particle physics; and

(*b*) that events in high-level sciences can be reduced on the basis of a one-for-one correspondence to events and hence laws appropriate to the lower-level science; ultimately, therefore, that physical laws can be derived which will subsume and explain sociology.

To appreciate the significance of reductionism as a philosophy, it is necessary also to recognise that reductionism as an experimental approach has been at the heart of the scientific method ever since the emergence of modern physics with Galileo and Newton. As an experimental method, reductionism is merely a procedure for explaining the properties of simplified, model systems, of holding all parameters except one constant, and varying that systematically. This makes the experimental problems under study more approachable; it has been the key to the success of the 'biological revolution' over the past two decades, and as a tool it is unchallenged. Problems only arise when the tool is elevated into a philosophical principle, so that it is ignored that, for a complete explanation of an event or a process, it must be taken out of the vacuum into which reductionism plunges it and replaced in the bustle of the real world with which it is, in actuality, in constant interaction.[1]

At one level, it is the very success of reductionism as a tool in the biological revolution, in the unravelling of the genetic code and exploring the chemistry of the cell, which has led to the ready way in which its philosophical premises too have become accepted. Thus we find molecular biologists such as Jacques Monod, author of *Chance and Necessity*, arguing that in the long run all of biology, and hence all 'higher' sciences, are to be derived from a study of the properties of the macromolecules of which the cell is composed (such as DNA) and their interactions, and may be best understood by studying the chemistry and organisation of the intestinal bacteria *Escherichia Coli* or — even more reduced — a bacteriophage, the virus which preys upon it.

Exposing reductionist ideology has become more complicated in that the modes of thinking of reductionism have become so dominant in recent years that they have come to constitute what may almost be described as an ideology of science itself, which claims that reductionism has universalistic importance, superseding all other forms of knowledge. The ideology of reductionism is thus positivist. But it also has ethical overtones, claiming that the scientific rationality it represents provides rules for the proper conduct of human society. The only true goal for humanity becomes, in this view, the systematic incorporation of all aspects of human existence into a framework provided by 'the laws of physics': the rationality and objectivity of reductionism replace all else; they provide their own guide to human progress. Science, a social product, becomes both the goal and the method for all society.

The opposition to reductionism comes primarily from dialectical materialism, which argues that, whilst events at any given hierarchical level represented by the different sciences must correspond to events at higher or lower levels, they cannot be reduced, by the application of causal laws or relations, to lower ones: biology cannot be invoked to explain away sociology; instead there will be a dialectical interaction between them. Thus, despite the vicissitudes of Marxism in the Soviet Union, there has been a constant attempt to maintain a non-reductionist science, which has been particularly successful in the case of neuro-biology.

But why is all this not merely a superstructural squabble about epistemology, without real relevance for serious political struggle or major ideological thrust? The answer to this lies in the present situation of capitalism – the particular strengths and weaknesses which have made it both necessary and possible to recruit biologistic biology as the generator of ideologies and technologies in its struggle for survival. To follow this we must look at the present role of biology and biologism.

So far as technologies are concerned, the major role of the new biology (outside of agriculture and some areas of medicine) is firmly locked into the processes of social control. The new needs which have generated the technologies and their attendant ideologies can be located in the changing nature of imperialist struggle abroad and class struggle at home. The characteristic mode of warfare of the latter half of the twentieth century has been anti-imperialist, guerilla struggle. Such struggles of national liberation essentially involve a people at war against an army (the change is symbolised by the statistic that in the First World War, 90 per cent of the deaths were of soldiers; in the Indo-china war of the 1960s and 1970s, at least 90 per cent of the deaths have been of civilians). In such struggles the highly mechanised imperialist army, equipped with devices generated by its physicists and engineers, has been outmanoeuvred by the guerillas, the fish in their pond of peasants. What is more, such struggles are no longer confined to the Third World, but take place within the metropolitan countries themselves; in Britain and the United States, urban guerilla war has become a characteristic feature of the landscape.

Physics has its contribution to such struggles in the form of fragmen-tation weapons, night-vision detectors, electronic sensors and com-puters, but biological methods become of increasing importance; the chemical crop destruction campaign in Vietnam was an example, and at home, where property is sacrosanct and large-scale damage unacceptable the pressure to develop methods for people control and manipulation, both on a general, population basis and aimed at specific individuals, has become very strong. The emergence of these technologies is heavily dependent upon the area of interface between biology, the social sciences and the so-called behavioural sciences – the field best known as 'neurobiology' — which has generated, and is generating, developments based on the extensive use of drugs on a population basis, behaviour modification techniques and the — widely resisted — use of psycho-surgery and brain and behaviour modification by electrical brain stimulation.

At the same time, the second, directly ideological, role of biologism as a legitimator of the social order has shown a massive resurgence. This is not the first time that biologism has played such a role. In the nine-teenth century, bourgeois theorists adapted evolutionary ideas to legitimise the capitalist mode of production and its consequent social relations as corresponding to the inevitable workings out of the 'iron laws of biology'. Both class structure at home and imperialist expansion abroad were justified under the name of Social Darwinism. In the present period of social crisis, where the more traditional legitimating ideologies of capitalism have been exposed and weakened, if not destroyed, the importance of biologism is of again 'proving' that capita-lism and imperialism derive from 'man's innate aggressiveness', that all human experience can be subsumed into categories of stimulus and response, reward and punishment, and that individuals' success or failure in a competitive society, their capacity to revolt against the state, is a result of a flaw in their chemistry or brain structure. This 'proof' both justifies the oppression and, by opposing their struggles with scientific rationality, devalues, divides and demoralises the oppressed. It is this ideological role, as much as its attendant technologies, which forms the present-day importance of biologism not merely at the superstructural level but in every present dimension of struggle.

In order to combat both the technologies and the ideology of biological reductionism, we have to examine its claims. In doing so, we must recognise that ideology, posing as science, itself creates paradig-matic frameworks within which endless apparently 'objective' research is done and learned journals and popular books produced. Particular technologies of oppression — whether of behaviour modification through conditioning, the apparatus of psychometrics or chemical brain manipulation — are frequently derived from a strange bastardisation of science and pseudo-science; their rationale is established within what is apparently a scientific, but in reality an ideological framework. As a result, particular technologies often appear merely as abuses of an otherwise value-free science, or are regarded as symbols of the inevit-ably oppressive role of a scientific rationality from which the only escape is retreat into irrationality.

Marxist analysis rejects both these partial accounts; we must instead link the technologies directly with the ideology which sponsors them, and show their coherence and social function. By analysing both ideologies and technologies as the products of a re-emergent biologism, we do not mean to imply that the technologies do not 'work' — drugs or psychosurgery will indeed 'pacify' individuals, even if they do so by reducing them to cabbages. This is the peculiar force of biologism: that ideology, occupying the terrain of the natural sciences, may indeed generate effective oppressive technologies. What we are concerned with is to show the underlying ideological coherence of the many forms of biologism which now compete with biology within the area of neuro-biology. To do this, in the following sections, we describe several different forms which biologism takes, and the attendant technologies which flow from them. For analysis, we separate the forms into varying types of reductionism, from molecular and genetic reductionism, through evolutionary reductionism to behaviourism. One whole area, that of scientific racism, a form of genetic reductionism, warrants an entire separate chapter.

## MOLECULAR REDUCTIONISM

Molecular reductionism can be seen at its sharpest in the explanation of madness. What is the cause of schizophrenia? Is it to be seen — as the school of 'orthomolecular psychiatry' would argue — in the absence of certain key chemicals in the brain or in the presence of abnormal metabolites due to genetic disorders? If so, treatment is to be found by dietary modification or the development of drugs which antagonise in some way the abnormal metabolites. Following the lead of such individuals as Osmond and Smythies[2] this school argues strongly that there is an organic, brain-located *cause* for the individual's response. This belief has a long history, for at all stages in the development of biochemistry the fashionable molecule of the moment has tended to be implicated as the cause of schizophrenia, from the amino acid glutamate in the 1950s, through an abnormality of energy metabolism in the 1960s to today's attention to the problem of the sugar galactose in the diet. Whatever the proximate biochemical cause, there is on this thesis an underlying genetic defect, a *propensity to be* schizophrenic. Although the original idea of the 'illness' as due to a defect in a single gene is now relatively disfavoured by comparison with more complex multiple gene effects, over the last decades all the classical apparatus of the clinical biochemists, in their search for bizarre substances indicative of abnormality in schizophrenic brain, blood, urine or sweat, and of the human population geneticists with their hunt for identical and non-identical twins and heritability estimates, have been employed, with singularly little success, to track down the offending chemical.[3]

The ideological components within the reductionist paradigm are apparent; the inborn view of schizophrenia at once refuses to admit criticism of social structures, such as the family* and alienated work forms, whilst at the same time encouraging a manipulative view of treatment. This is even more apparent when we look at the respective analyses, biochemical or social, as applied to the affective disorders such as depression. Those who argue a biochemical cause of depression, such as the psychiatrist Sargent,[4] look for treatment by way of anti-depressant drugs; treatment is effective if it adjusts the depressed individual (typically a woman post-natally or around menopause) back into an acceptable social role, such as that of the good housewife and mother. The stability and appropriateness of the social order is taken as a natural given in this situation, and the job of the psychopharmacologist and clinician is to chemically fit people to it; it is not surprising to learn that 50 million patients were given chlorpromazine within the first decade of its use, or that 12 million barbiturate and 16 million tranquilliser prescriptions are issued a year in Britain. Note that we do not argue that the drugs themselves do not 'work' in the sense that they affect an individual's responses and performance, often, though not always, in the predicted and hoped-for (by the clinician) direction. However, even here there are some difficulties, because one consequence of the reductionist mode of thought is that drugs are supposed to have a single site of action; effects other than those sought for are seen as 'side-effects' to be eliminated. The complexity of drug-behaviour interactions, which has been so revealingly brought out even in the context of an apparently much less controversial agent, L-Dopa, in the treatment of an apparently straightforward motor disorder, Parkinsonism, tends to be dismissed by the reductionists.

But this is only a relatively marginal point; the crucial issue is that the ideology which reflects itself in a reductionist model of the cause of

---

*However a causal theory of madness which locates the problem exclusively within the family (as did R. D. Laing and A. Esteson in *Sanity, Madness and the Family*, Harmondsworth, Penguin, 1970) while opening the way to criticism of one particular oppressive social structure, limits itself by excluding both other social forms and human biology itself from theoretical consideration.

schizophrenia and depression, and upon which is based a vast research and development programme at all levels, from the universities to the drug houses, also finds expression in an output at the social level, which essentially regards individuals as objects, to be manipulated into required social patterns. Compare the average clinical research paper's description of the behaviour of patients exposed to given agents with the multi-levelled account of the patients to whom the doctor Oliver Sacks gave L-Dopa, in his book *Awakenings*.[5] For each of his twenty patients Sacks provides an account of the workings of L-Dopa, in terms of both minutely observed behaviour associated with different levels and occasions of giving the drug, and also of the personal histories and present relationships of the patients. Throughout there is a constant emphasis on the integration of all the effects of the drug into the complex individual situation. Sacks' model is clearly a dialectical one: truly scientific rather than ideological.

The search for a biological rationale for problems of the social order has reached new heights in recent years with the development of a new clinical concept that of 'minimal brain dysfunction' (MBD), which, an almost unrecognised category a few years back, was recently warranted a full symposium of the New York Academy of Sciences,[6] Minimal brain dysfunction is essentially a behaviour-defined syndrome; that is, the concept of brain dysfunction is invoked to explain a particular pattern of socially disapproved behaviour, although no brain abnormality can in fact be detected by physiological techniques. In general, minimal brain dysfunction is supposed to be a disease of childhood, and has derived from an extension of the concept of 'hyperkinesis', a disease state believed to characterise an overactive child. In Britain, there are estimated to be a few hundred children categorised as hyperkinetic. A large proportion of these are institutionalised; they are described as being unable to sit still without forcible restraint. In the United States the diagnosis of hyperkinesis has become much broader, to cover a very large group of children who show 'behaviour problems' at school, being poor learners, inattentive in class, and disrespectful of authority. Amongst the diagnostic signs for minimal brain dysfunction are, according to Wender, being 'aggressive socially . . . playing with younger children, and, if a boy, with girls'. For all these patterns of behaviour, treatment with amphetamine, or its congener, Ritalin, is proposed. Indeed, Wender waxes euphoric about the effects of Ritalin treatment, going so far as to claim that minimal brain dysfunction children may be regarded as suffering from 'hypoamphetaminosis'.

Under the drug, 'children often begin to talk about and behave in a manner consistent with their parents formerly unheeded "oughts" and "shoulds" '. One bright eight-year old referred to d-amphetamines as his 'magic pills which make me into a good boy and make everybody like me'. The child turns from a 'whirling dervish' into being 'quiet, compliant' and with an 'improved class behaviour, group participation and attitudes to authority' under medication. What is more, Ritalin is cheaper than 'expensive, non-organic therapies'.[7] *

Small wonder that Ritalin is now prescribed, at a daily dose of 5 – 40 milligrams, and on the basis of school reports, to 250,000 American schoolchildren daily. Here indeed is a conspicuous success for a reductionist research and development programme that discards any alternative explanation for a child's inattentiveness in class or poor attitude to authority; not even other biological factors which may have apparently similar consequences, such as nutrition, are considered, still less that inattentiveness in class may reflect poor teaching or an irrelevant educational programme, or that disrespect for authority may represent a more socially appropriate response to oppression than does subservience.

Biochemistry is not the only brain discipline whose reductionism has both ideological and direct social significance. Physiology and anatomy have shown similar tendencies. Over recent years it has become increasingly apparent that electrical activity in the cells of particular brain regions is associated with specific behaviour patterns, so that, for instance, when certain nerve cells in the hypothalamus, a region deep in the interior of the brain, are electrically stimulated by implanted electrodes in the cat or rat, then, depending on the particular cells stimulated, the animal shows hunger, thirst, satiety, anger, fear, sexual arousal or pleasure. Surgical removal of these regions is associated with the reciprocal behavioural effect to that of cellular stimulation. The reductionist interpretation of these experiments is that the firing of particular cells in the hypothalamus *causes* anger, sexual arousal, and so on, and, as with the biochemists, the social technologies which have emerged, notably in the hands of Delgado in the United States and in Spain[8] have been human experimentation in which schizophrenics and 'low IQ' patients have had electrodes permanently implanted in the

---

*In Britain, amphetamine prescription is generally discouraged by the BMA. Connoisseurs of British children's comics will also note the strong resemblance of Wender's disapproved-of hyperkinetic child to the weekly *Beano* hero Billy Whizz.

E

hypothalamus, remotely radio controlled by the doctor/experimenter. Passing current through the electrodes is associated with sharp mood changes in the patients. Once again, the patient's anger, arousal, and so on is seen as a consequence of the functioning of particular brain cells; the cells can be manipulated and so the patient can be manipulated, irrespective of the external circumstances which might be expected to affect the individual's mood. According to Delgado, implanted electrode studies and their utilisation in practice can be expected to develop substantially in the next few years.

Still more revealing is the recent growth in popularity of techniques for the removal or destruction of particular brain regions — psychosurgery as it is called — in the United States, and also in Britain, Japan and other countries.[9] The protagonists of these techniques argue that particular behavioural patterns are associated with malfunction or hyperfunction of particular brain regions, so that the appropriate medical strategy is the removal of these regions, a surgical approach which is a modification of the old pre-frontal lobotomy popular for use of schizophrenics in the early 1950s, but more recently a relatively declining treatment. Psychosurgical operations have recently been rapidly increasing in number in both Britain and the United States.

Increased knowledge of the hypothalamic centres and related regions of the limbic system (a part of the brain associated with fear, anger and similar emotional responses) has led to a considerable ramification of these techniques. Surgical removal of such brain regions has been both proposed and practised to deal with individuals suffering from 'behaviour problems' without any obvious 'organic' brain dysfunction. Such psychosurgery is intended as a pacifier, producing better-adjusted individuals, easier to maintain in institutions or at home. In the United States the commonest groups of patients are claimed to be working-class blacks and women. A book by two psychosurgeons, Vernon Mark and Frank Ervin,[10] has drawn on the circumstances of the revolt in American cities to query whether there may not be brain abnormalities present in ghetto militants, to be cured by surgery. Their estimates argue that between 5 and 10 per cent of Americans might be candidates for such treatment.

Nor is this discussion purely theoretical; psychosurgical research has been supported by law-enforcement agencies in the United States. Whilst in a Detroit court case in the early 1970s, proposed brain surgery on a prisoner was ruled illegal despite his consent, the number of actual operations conducted apparently continues to increase. An indication of the candidates for such operations is provided by an exchange of correspondence between the Director of Corrections, Human Relations Agency* (sic) Sacramento, and the Director of Hospitals and Clinics, University of California Medical Center, in 1971.[11] The Director asks for a clinical investigation of selected prison inmates 'who have shown aggressive, destructive behaviour, possibly as a result of severe neurological disease' to conduct 'surgical and diagnostic procedures . . . to locate centers in the brain which may have been previously damaged and which could serve as the focus for episodes of violent behaviour', for subsequent surgical removal.

An accompanying letter describes a possible candidate for such treatment, whose infractions whilst in prison include problems of 'respect towards officials', 'refusal to work' and 'militancy'; he had to be transferred prisons because of 'his sophistication . . . he had to be warned several times . . . to cease his practicing and teaching Karate and judo. He was transferred . . . for increasing militancy, leadership ability and outspoken hatred for the white society . . . he was identified as one of several leaders in the work strike of April 1971. . . . Also evident at approximately the same time was an avalanche of revolutionary reading material'. To which request, the Director of Hospitals and Clinics replies, agreeing to provide the treatment, including electrode implantation 'on a regular cost basis. At the present time this would amount to approximately $1000 per patient for a seven day stay.'

Clearly such cases are those to whom the label 'these animals are dangerous; when attacked they bite' might well be attached. Once again, the reductionist slogan is the reverse of that painted on the Oxford College wall: 'do not adjust your mind; there is a fault in reality'.

In such examples of molecular reductionism we see an amalgam of all those features of the ideological penetration of science discussed above. The research paradigms not merely dictate the experimental operations conducted, such as the search for abnormal metabolites or particular 'centres' in the brain, but have an ideological significance which lies both in determining scientific directions and in providing a powerful scientific rationale for particular social interests. But, not only do these paradigms provide ideological support for the existing social order (it is your brain that is at fault if you are disaffected), they also provide a set of social technologies which help maintain precisely the same social order.

---

*What need for a Ministry of Love?

GENETIC DETERMINISM

Genetic determinism represents a particular paradigm within the broad framework of molecular reductionism. Its research programme is based upon the premise that all human behavioural characters can be analysed as representing the algebraic sum of two components: a contribution from genetics and a contribution from the environment, with a further separable balancing item for interaction included. From this premise follows the belief that experiments can be devised to answer the question: 'how much does environment, and how much does heredity, contribute to differences in behaviour between individuals or between populations'. Note that such a question, with its implicit claim that behavioural characters can be teased apart and reduced into elemental components which summate, is itself archetypally reductionist.

While genetic analysis of particular behaviour traits has been attempted with non-human animals, it is with humans that most of the work is concerned. Heritability studies have been performed on traits such as schizophrenia and other mental disorders and there have been attempts to explore the genetic basis of criminality (for instance the attention paid to the proposed relationship between a particular genetic defect found in certain males, the XYY chromosomal abnormality, and a propensity to violent crime). This relationship was first canvassed some years ago in Britain, and had apparently fallen into disfavour,[12] but recently has emerged again as the subject of a vigorous research programme in Boston, Massachusetts, in which the parents of male children, screened and found to have the abnormal chromosome, are told at the child's birth that he may grow up abnormally!*.[13]

However, not only has there been a debate concerning the existence of genes for criminality, there has also been a re-emergence of the claim that there are genes for low IQ. Because this is *the* outstanding example of biologism we have given the IQ racket a chapter to itself, but it must be seen as a key element within the general picture of a neurobiology penetrated by ideology.

EVOLUTIONARY REDUCTIONISM

Amongst the several reductionist strands present within genetic determinism, some perhaps belong more properly to that paradigm group we have classed as 'evolutionary reductionism'. The most clear-cut and prominent examples of this paradigm are provided by certain tendencies

---

*This programme was curtailed in 1975 following strong political action by the Boston Science for the People group.

within ethology, the study of the behaviour and social relations of animals studied as far as possible in a natural environment and un-restricted by laboratory conditions. The development of ethology, evolving as it did in reaction to the sterility of much laboratory psychology (see below) has certainly provided a new approach to an understanding both of patterns of behaviour and of relationships bet-ween individuals of a species, which has enriched the understanding of the complexities of social behaviour. However, it has also manifest within it some very obvious and vulgarly ideological reductionist models, typified by, for instance, Desmond Morris' *Naked Ape*,[14] in which he argues that human conduct is most fruitfully interpreted, predicted and controlled in the light of studies of other primates. Whilst Morris' more extreme books, or for that matter Robert Ardrey's *Territorial Imperative*[15] — which makes claims for the innate aggressive-ness of human beings and their urge to possess 'territory' — are by and large deplored by professional ethologists as being orientated towards the lay rather than the professional audience, they are none the less influential in determining research both within ethology and in neigh-bouring areas, and they have generated hosts of anthropological and sociological camp followers who busy themselves in demonstrating how today's managerial capitalist society is the direct biological (inevitable) descendent from humanity's hunting past and non-human ancestry.[16] What is particularly apparent in these publicist accounts of ethology is the clarity with which they articulate some of the central dogmas of ethological authority. Thus the innate aggressiveness of humans is claimed directly by experimentalists like Lorenz and Eibl-Eibesfeld,[17] so whilst Ardrey's exposition of territoriality in man derives sustenance from research studies on territoriality in red grouse on Scottish moors, extrapolated to the human world.[18]

Reductionist paradigms, in which a mode of operation becomes elevated by some invisible hand into a principle, is like goal displace-ment in organisations; a kind of explanation displacement occurs so that research which may provide an elegant account of certain aspects of animal behaviour is displaced into a total account of the whole human condition.* Scarcely surprisingly, if humans are interpreted as ill-suppressed bundles of aggressive instincts, the formulation for social policy relates to control rather than liberation. Thus an ethologically based legitimation for conserving the social order is provided by the dominance hierarchy ('pecking order') studies; stratification is not

---

*Now graced with a newly fashionable reductionist title: 'sociobiology'.

associated with specific societies and cultures, but reflects a genetically laid-down necessity. The limitations of this particular type of ethological approach have been criticised by for instance, Patrick Bateson,[19] who has pointed out that not only do studies of pecking orders and dominance hierarchies relate merely to particular species examined under particular conditions, but, in addition, even within a group, the pecking order itself is not rigidly ordained but relates rather precisely to a particular type of experimental situation; in other situations, quite different orders may obtain, so that different hierarchies may be apparent between for example, eating activities and sexual activities.

Reductionist analyses of hierarchies, even in non-human species, must be replaced by dialectical ones which take into account the environmental circumstances of individuals and their past experience. But a reductionist ethology is one which, by definition, appropriates a set of linear and pared-down analyses of particular situations, and therefore is far more prone to extract out from the richness of the experimental data the simplistic and linear concept of a pecking order or a dominance hierarchy. In so far as the social and political beliefs of such ethologists are apparent from their writings, there are few areas of contemporary neurobiology in which ideology stands out so sharply as in the work of ethologists such as Konrad Lorenz, in his time a paid-up member of the Nazi Party, now a Nobel Laureate and author of his own tract for the times on human survival, *Civilized Man's 8 Deadly Sins* (sic).[20]

In this respect, reductionist ethology in the 1960s and 1970s has played, and is playing, the same role that Darwinism in the form of Social Darwinism played in the nineteenth century.[21] Then, Victorian capitalism was interpreted as obeying the iron laws of biology; the struggle for existence and the survival of the fittest demanded a *laissez-faire* economy at home and legitimated imperialism and colonialism abroad. Today, managerial capitalism, bureaucracies and social stratification, and social conflict of all types, from football hooliganism through class war to struggles of national liberation and wars between nation-states, are seen as the inevitable results of human evolution from the primate. Such explanations ignore the entirely new dimensions to human behaviour generated by the human capacity for communication, social existence and, above all, production. Economic and sociological accounts of conflict or social structures thus become diminished to the working out of the evolutionary imperative.

## BEHAVIOUR REDUCTIONISM

Our final example is drawn from the powerful school of psychology, that known as 'behaviourism'; here, the framework into which all human behaviour is to be reduced is that of reward and punishment, the so-called 'contingencies of reinforcement'. Behaviourist theory is one which is simultaneously extremely environmentalist and highly reductionist. It takes almost as a tenet of faith that all aspects of animal or human behaviour can be, and are, shaped by means of particular combinations of rewarding or aversive stimuli. At the same time, however, it claims to be able to reduce all aspects of human activity to a system of 'emitted behaviours'. What is important to behaviourism is what is measurable; events which occur within the brain and which are unobservable (intervening variables) are of little importance. The animal model for human behaviour favoured by the behaviourist is that of a rat or pigeon in a box provided with a lever it can press for reinforcement; indeed the key behaviourist concept is that of 'reward'. This approach to human behaviour is a classic category reductionism, where all aspects of human activity, from the writing of an academic paper, through the factory production line to altruistic self-sacrifice in war or struggle,* are defined as behaviours emitted in mechanistic response to past patterns of reinforcement for the individual. The behaviourist school is sharply distinguished from other psychological paradigms, publishing its own journals and regarding as its mentor B. F. Skinner, and it is therefore interesting to examine the behaviourist position on human behaviour as evinced by Skinner's book *Beyond Freedom and Dignity*,[22] in which he argues that all human activity is embraced within its concepts. This type of reductionism is at its worst when Skinner considers the relationship of culture to individuals, serving to control and manipulate them. He cannot see that the contradictions between individuals are themselves a part of and contained within the over-all structure of society, that it is not culture as a reified abstract which controls individuals, but that culture is a product of competitive classes and groups within society. Parents and teachers manipulate and control children, as Skinner points out; but it is ignored that these parents and teachers have themselves in their turn been manipulated and controlled.

---

*By contrast, sociobiology seeks a 'genetic' explanation of altruism on the grounds that self-sacrifice by the individual in defence of relatives will preserve some of that individual's genes for posterity — hence altruism shows a selective advantage — provided you only save your kin!

Because of this, despite Skinner's emphasis on the possibility of designing a culture, there is an ahistoric, static quality about his concept of society. Nowhere does he present a vision of a future culture; instead he emphasises the 'ethical neutrality' of his techniques, applicable presumably equally to fascism, liberal democracy or socialism. Simultaneously he makes the strange error of claiming that 'no theory changes what it is a theory about'. Yet the remarkable thing about humans and their society is that they *are* changed by theories, precisely because theories modify consciousness. In fact, because Skinner's ahistoric concept carries conviction only within the atmosphere engendered by a society of the sort Marcuse characterised as one of repressive tolerance, Skinner's position is irreconcilably conservative, and its emphasis on reward as the unifying concept for describing human behaviour is deeply ideological.

## COMBATTING BIOLOGISM

Throughout this account of reductionism in neurobiology, we have attempted to show both that reductionism is more than merely 'bad science', in the Anglo-Saxon sense of being poor theory or ill-conceived experiments, but that it is bad science *because* it is ideological, that is its research programme and organising paradigms are permeated with those 'ruling ideas' which express class interest, and that the technologies that they generate are essentially defensive of that class interest, serving to protect it both physically, by manipulating and pacifying would-be protesters, and ideologically, by providing an apparent biological justification for the social order.

How can biologism be combatted? There are those, particularly within the alternative culture, who respond to its oppression by turning away in distaste not only from the technologies but also from the 'science' they see as generating them. In disgust with the inhumanity of science, the young Bohemian strata turns again towards the irrational as an explanation for human suffering and joy. As the risk of fascism grows in Europe the support of a seemingly innocent irrationalism unwittingly increases the dangers. As Horkheimer wrote during a previous wave of irrationalism, 'the philosophical dismissal of science is a comfort in private life, in society a lie'.[23]

The point is that, despite the loss of the terrain held by science to ideology, particularly in the crucial area of neurobiology, to abandon science along with scientism is a sure road to defeat, a way to ensure the strengthening of the very system which generates biologism. Nor is it enough merely to 'expose ideology' in the shelter of 'anti-bourgeois research' or cultured lectures to student audiences. Instead, the way forward must lie in linking the superstructural struggle with that in the work-place, the home and the streets. Powerful movements of resistance have sprung up and are growing against biologism's brutal pessimism. Campaigns of parents and schoolchildren have developed in Britain against, for instance, ESN-labelling, particularly of black children, and against Ritalin in the United States. The developing American prisoners' movement has used both agitational and legal forms of struggle against psychosurgery and the massive behaviour modification programmes which form part of the US 'law enforcement' strategy. The campaigns against racism in Britain and the United States have not been confined either to factory struggles or academic disputations of 'hereditarians versus environmentalists' but have identified scientific racism as one of the main enemies to be fought at all levels. The point is that such struggles, if they are to succeed, cannot ignore theory in the development of practice.

# 7

# Scientific Racism and Ideology: The IQ Racket from Galton to Jensen*

*Steven Rose*

The natural ability of which this book mainly treats is such as a modern European possesses in a much greater average share than the men of the lower races (Francis Galton, statistician and eugenicist, England, 1869).[1]

No rational man, cognizant of the facts [could deny that the Negro is inherently inferior]. It is simply incredible that, when all his disabilities are removed, and our prognathous relative has a fair field and no favor, as well as no oppressor, he will be able to compete successfully with his bigger-brained and smaller-jawed rival, in a contest which is to be carried on by thoughts and not by bites (Thomas H. Huxley, England, evolutionist, 1896).[2]

The path of progress is strewn with the wreck of nations; traces are everywhere to be seen of the head tombs of inferior races . . . the stepping stones on which mankind has risen to the higher intellectual and deeper emotional life of today (Karl Pearson, England, biometrician and eugenicist, 1900).[3]

In the actual race of life, which is not to get ahead, but to get ahead of somebody, the chief determining factor is heredity (Edward Thorndike, U.S. educational psychologist and eugenicist, 1927).[4]

*This paper draws on an earlier one, 'Science, Racism and Ideology', written jointly with John Hambley and Jeff Haywood, which appeared in *The Socialist Register* (1973). Much help was given by members of the Brain Research Group and the Campaign on Racism, IQ and the Class Society.

As compared with the European races, the Negroes certainly lack foresight. In general, a Negro is not inclined to work hard in the present in order to provide for wellbeing in a distant future. The Negro is more strongly influenced than Europeans by the immediate impression of the senses, and is therefore much more strongly attracted by gewgaws. According to the nature of his present experience, he vacillates between a cheerful indifference and a hopeless depression. . . . In the Mongol, it is above all the heredity factors tending to promote an aptitude for social life which are well developed; but on the whole he has more capacity for imitation than for invention . . . every Chinaman lies, even when he seems unlikely to gain anything by it. . . . The intellectual gifts of the Alpine race are in general, notably inferior to those of the [Nordic]. . . . The Mediterranean . . . occupies an intermediate position between the Nordic and the Negro. . . . The Jews . . . are . . . distinguished by a peculiar aptitude for trade and commerce . . . [they] have been selected for an instinctive desire not to look singular . . . we have to do with the mimicry which is so fundamentally observed wherever a living creature gains advantages in the struggle for existence by acquiring a resemblance to some other organism (Fritz Lenz, Germany, human geneticist, 1931).[5]

By 'racism' is meant any claim of the natural superiority of one identifiable human population, group or race to another.

By 'scientific racism' is meant the attempt to use the language and some of the techniques of science in support of theories or contentions that particular human groups or populations are innately superior to others in terms of intelligence, 'civilisation' or other socially defined attributes. There is nothing new in the existence of racism; its history predates contemporary genetics, biometrics and evolutionary theory. However, it was only with the development of the biological and human sciences of the nineteenth century that racist ideas abandoned their buttressing of God and 'divine right', and began instead to turn to biology for ideological support.

Whilst racism takes many forms, and scientific racism has a wide range of choice as to areas for an attempt to prove superiority and inferiority, there is no doubt that at present – and for much of the last hundred years – its attention has turned on the question of intelligence. For more than twenty years, from the 1940s to the late 1960s, scientific racism lay dormant; many believed it dead; the coffin of Nazi pseudogenetics had been well nailed down, the myths of the 1930s apparently destroyed, and those who had helped push the corpse in and mounted guard over it felt able to close down their picket line and go elsewhere. Yet in the late 1960s and increasingly vociferously since, there has been a vigorous attempt to revive the corpse, and to re-open that peculiar alliance of hereditarian argument and educational psychology that has characterised eugenicist and racist thought at least since Galton's day. The extraordinary thing about the whole discussion is, as we shall see, that there is a curiously static quality about it; with a few changes in terminology, the challenge and counter-challenge could have been made in Thorndike's or Lenz's period – or Pearson and Galton's.

## THE RE-EMERGENCE OF SCIENTIFIC RACISM

The chronology of this re-emergence is easy to document. In 1969 the *Harvard Educational Review* carried a long article by a relatively little-known educational psychologist from California, entitled 'How much can we boost IQ and Scholastic Achievement?'.[6] The author, Arthur Jensen, concluded that the reasons for the apparent failure of certain 'compensatory education' programmes in the United States lay, not in the inadequacies of the teaching itself, nor in the social structures which generate poverty in the United States, but in the innate, genetic inferiority in intelligence of the groups concerned – mainly Blacks. As he put it: 'There are intelligence genes, which are found in populations in different proportions, somewhat like the distribution of blood types. The number of intelligence genes seems lower, overall, in the black population than in the white'.

The article created an immediate furore. In the United States, Jensen's lead was followed by a number of others, notably Richard Herrnstein,[7] who extended the analysis to claim that most of the determinants of class society were also due to genetic differences, and William Shockley (Nobel prize winner in physics for work on the transistor), who derived the 'logical' policy conclusions by noting that, as Blacks and the White working class tend to have larger families than the White middle classes, it followed that the national intelligence was declining. Shockley's policy recommendation was for a programme of cash inducements for sterilisation, linked by a sliding scale to the sterilisee's IQ score.[8] For Shockley: 'Nature has colour-coded groups of individuals so that statistically reliable predictions of their adaptability to intellectually rewarding and effective lives can easily be made and profitably be used by the pragmatic man in the street.'[9] When

Jensen's work was cited by US segregationists in support of their school policy, he did not repudiate them.[10] When student protest mounted against the three men, the press claimed they were being martyred for their beliefs.

The issue was brought to Britain in 1970, when Jensen spoke at a meeting organised by the Society for Social Responsibility in Science at Cambridge.[11] Five opponents of his views also spoke and the press claimed the meeting had been stacked against him; that he was being persecuted for 'heresy'. The following year, a book endorsing Jensen's line, and avowedly a 'response' to the Cambridge meeting; *Race, Intelligence and Education*,[12] was written by Hans Eysenck, Professor of Psychology at the Maudsley Institute of Psychiatry in London, and a past teacher of Jensen.

The book was written for a popular audience; in his text, Eysenck first carefully disclaimed any racist intent, pointing to his own experience as a Jew in Nazi Germany and then argued a detailed case in defence of the 'Jensenite Heresy', that Blacks (and the Irish and the working class) were genetically inferior in intelligence to Whites (and the English and the middle class). His conclusion is summed up as:

It seems certain that whenever blacks and whites are compared with respect to IQ, obvious differences in socio-economic status, education and similar factors do not affect the observed inferiority of the blacks very much . . . this inferiority . . . cannot be argued away as being due to lack of motivation.

The book achieved considerable notoriety and triggered a number of replies.[13]

By this time, the controversy had long since ceased to be (if it ever was) 'academic'. Eysenck's work was used to 'explain' the high percentage of West Indian children in ESN schools and as a further justification for the education premises of the authors of the so-called 'Black Papers' which advocated a return to more selective education. To all the other arguments for racism in education, in the trade unions and elsewhere, could be added 'but they're stupider — scientists have proved it'. Yet there was further advantage to be obtained from the martyr stance. When Leeds University offered Shockley an honorary degree in 1973 for his transistor work, and then withdrew it when they were made aware of his more recent research proposals and policy recommendations, he was not slow to seize the publicity possibilities. Nor was Eysenck when, having been invited to speak at the London School of Economics by social science students, he was prevented from doing so and had his glasses broken in the ensuing melee. All he wanted, after all, as a *Guardian* leader put it, was some peace to 'advance research in the field of genetics'.[14] Facts, even unpalatable facts, must be faced; we must pursue scientific truth with objectivity, wherever it may lead. The apotheosis of the 'new Galileos',[15] brave and persecuted scientists, had come. No one bothered to point out that the ideas and research which Eysenck, Herrnstein, Jensen and Shockley were purveying, though trimmed out in modern style and appropriately sophisticated, were not new at all, but very old and long discarded.

Whether Eysenck, Jensen and their followers are or are not racists is of no relevance here. We must first examine the claims to scientificity that the conclusions they draw possess; that is, the 'facts' and 'theories' they claim support their views. Second, we must look at the social and political resonance of these views; that is, the history of the IQ testing and eugenics movement. Finally, we need to explore the social framework within which the views are drawn into prominence — that is, the ideological role that they fulfil.

## IQ AND GENETICS: SOCIAL PSYCHOLOGY AND BIOLOGY

To focus our discussion on the nature of the Eysenck/Jensen claims, we can begin by summarising them as a set of propositions, each of which we can then examine. They run as follows:

(1) There is a thing, intelligence, which IQ tests measure;*

(2) The working class, the Irish, Blacks and Mexican Americans, score lower on IQ tests than the middle class, the English and White Americans;

(3) Studies on the heritability of IQ within the White population, based largely on the evidence of identical (monozygotic) twins reared apart, suggest that 80 per cent of the variance between individuals can be parcelled out as genetic, 20 per cent as environmental;

(4) For the purposes of applying these calculations to social groups, Blacks, Mexican Americans, and so on can be regarded as representing biologically defined as well as socially defined races; and

(5) The differences between these groups are larger than can be accounted for by the 'environmental' factor, and hence are genetically based.

---

*Although on occasion Eysenck likes to refer to intelligence as a *concept* rather than a thing, it is difficult to see how a concept can have its heritability estimated or its habitation sought.[16]

From this set of propositions certain policy conclusions may be allowed to flow, varying from the circumspect suggestions that different types of intellectual apparatus require different types of eucation made in Jensen's original paper through to Herrnstein's genetic meritocracy and the cheerful cash-incentive eugenic sterilisation programme of Shockley.

All of these propositions can be demonstrated to be fallacious, with the exception of the second, which can be shown to be merely an artefact of the way the tests are constructed, and it is important to go over the arguments which show them to be so. Before doing so, however, we may note that they are all aspects of reductionist thinking, a general tendency to seek biological explanations for social issues, an appeal to the apparent scientificity of biological data which, although none of the men concerned is a biologist, is frequently used with debating skill to score points off non-biologically orientated educationalists or psychologists. This biological reductionism assumes that biological differences *cause* social differences. Further, any aspect of human behaviour, it is claimed, must essentially be the result of two separable components: genetics and environment. Because these components may interact, a third term is therefore also introduced to account for this interaction. It is this belief in the *separability* of component parts of a complex which even begins to allow the deceptively, fatally naive question: 'what proportion of the difference in performance between two individuals or groups is due to genetics and what to environment?' to be asked at all. As we will show, the very question is scientifically meaningless (a more extended treatment of the arguments advanced here will be found in reference 17).

IQ AND INTELLIGENCE

Jensen's and Eysenck's position is that, although they are imperfect measuring tools, none the less IQ tests do measure a unitary biological reality, a thing sometimes referred to as Spearman's *g*, a 'general intelligence factor'. Eysenck even goes further; he speaks of the 'habitation' of intelligence within the brain, as if it were a homunculus, and seeks to correlate IQ scores with EEG patterns almost in the manner of a nineteenth-century phrenologist. Neurobiologists (within which are included neuroanatomists, neurophysiologists, neurochemists and experimental psychologists) find such a model surprisingly old-fashioned; the components which make up an animal's — or a human's — performance in any type of test or learning situation are manyfold and interacting — they include attention and arousal, stress levels and perception, sensitivity to the features of the test situation, and so on. No one factor can be isolated as 'intelligence' in this sense, any more than can 'philoprogenitiveness' or 'mysogeny'. For instance, one strain of animal may perform better than another in a maze in which the punishment for incorrect choice is electric shock, not because they are more 'intelligent' but because their footpads are more sensitive to pain and therefore the need to escape is more urgent. It is well known that after generations of inbreeding strains of rats or mice which are 'maze-bright' or 'maze-dull' can be produced. Yet the bright and dull strains, when given a different type of task than the particular mazes on which the differentiation was based, may not show any such differences. Where is the $g$ factor here?

Yet the proponents of the genetic basis of IQ seem to have little doubt about what their tests measure; they often use the two terms, IQ and intelligence, interchangeably. There are several varieties of IQ test, all testing the ability to manipulate a combination of some or all of figures, numbers and words. Tests which rely on non-verbal skills and which present items not easily related to general knowledge are termed 'culture-free', and in theory they should be equally difficult for any person, no matter what their background. Yet one standard test, the Stanford–Binet, contains pictures of faces, all white, some obviously middle-class, others more battered by life, and asks 'which is the prettier?'. Other questions include: 'what is the thing for you to do when you have broken something that belongs to someone else?' — correct answers, according to the testing manual, involve 'restitution or apology or both; mere confession is not satisfactory'. To the question 'what is the thing to do if another boy (girl, person) hits you without meaning to?' the 'only satisfactory responses are those which suggest excusing or overlooking the act' (for example, 'Tell them they never meant to do it.' Incorrect is, for example, 'I would hit them back.'). In general IQ scores correlate highly with scholastic achievements, and their predictive value in this area is one use to which they have been put. Children have been screened for scholastic aptitude, segregated on the basis of this as measured by IQ and then given educations to fit their abilities. Because scholastic achievement plays a large part in the choice of occupation, IQ scores, inevitably, have a reasonable correlation with socio-economic status. High IQ characteristics as revealed by the test are thus middle-class or upper-class background or values, a positive attitude to school and teachers, respect for property, patriotism and docility and, in general, an acceptance of bourgeois values.[18]

The IQ test is thus essentially a social construct (originally, the ratio between a given child's 'mental age' as revealed by test scores and his or her chronological age) and is manipulated deliberately in order to provide a particular normal distribution of scores. The curve is symmetrical (with a small tail) about the median for the population of 100. But this symmetry is deliberately achieved by careful choice of test items; those which disturb the symmetry are removed, and tests which do not give a 'normal' (more or less) distribution for the test population are discarded. Thus, in early versions of the tests, males and females scored differently on certain items (the females scored higher). The tests were modified, and on the revised tests within the White population, the sexes now score more or less identically. Which measures the biological 'reality' – the test with or without the differential scoring items? Yet it is on such tests, on which American Blacks average a score of 85 compared with 100 for the White population, that the argument hinges.

It is possible to devise tests on which the working class score higher than the middle class, or Blacks higher than Whites. However, such tests are discarded, and the fact of the Blacks' better score is actually used, for instance, by Jensen, to claim that it represents a lower-order intellectual skill (for example, Jensen invents a hierarchy of so-called 'level-I' and 'level-II' skills, the 'lower-order' skills being more routine, the higher more 'creative'. Level-I skills are those the Blacks score more highly on). A similar explaining away is used in respect of the observations that in certain cultures, Black babies tend to be more advanced than White babies in terms of sensori-motor co-ordination; here Eysenck and Jensen claim that this illustrates a so-called 'biological law' that animals of lower final range of learning potential mature faster than those of a higher final range.[19] The dubious validity of this proposition is demonstrated clearly by comparison of two similar species, rat and guinea pig. The rat is born immature – blind, naked and with much of its brain development still to come. The guinea pig by comparison is much further developed at birth; its eyes are open, it has a coat of hair, can run well and its brain is much closer to the adult. Yet one may doubt whether even Eysenck and Jensen would wish to maintain that the rat has, in parallel to its relative prematurity, a substantially greater learning potential than the guinea pig. Perhaps psychologists should be more careful in arguing about 'biological laws'.

Even so-called 'culture-free' IQ tests cannot adequately compensate for the known effects of group differences in perception itself, both between geographically separated cultures (advanced urban industrial versus rural peasant) or between classes. An example is the study made by Lewis,[20] who argued that working-class children had to live in an environment in which they are subject to much more lies and misinformation ('noise') than middle-class children. He devised tests in which the testee had to devise a strategy despite a great deal of such misinformation, and compared a group from a local working-class cafe to a group from Mensa — the society for those with high IQs. The working-class group did considerably better. What does this prove? Merely that tests test what they are constructed to test. Standard IQ tests continue to test for performance regarded as appropriate to bourgeois standards.

Attempts to avoid the cultural problem in relation to Black/White differences include, for instance, doing IQ tests on Black and White groups matched for what is called 'socio-economic status' — that is, groups doing roughly the same type of job, of the same age and educational background, and so on. The differences in IQ persist, and it is claimed that this 'proves' the difference is biological. But to be part of a minority group in a majority culture which has enslaved your ancestors and discriminates against you on the grounds of your skin colour represents a cultural difference which cannot be equilibrated by naive manipulation of 'socio-economic status'.

Yet another problem arises in the use of IQ tests, however, which springs from the actual test situation itself. Tests do not represent the application of a neutral instrument, a test, by an objective tester, to a testee whose performance is being measured. Rather the results of a test are themselves the products of a three-way interaction between tester, test and testee. Whilst the contribution of the testee to this product is probably the most substantial, the other components of the interaction cannot be ignored. Such interactions manifest themselves at the level of labelling theory, in which, for example, teacher expectations of child performance may modify that performance, or in the reports that Black children may score better on IQ tests administered by a Black (or even by a computer!) than by a White.[21] Such interactions are ignored where intelligence is seen as a relatively fixed attribute, constant through life. Yet it is well worth emphasising a crucial point which has been largely brushed aside by those wishing to prove a 'genetic' theory. Although IQ in later life is relatively stable, before the age of eight,

IQ scores are very variable and have a very low correlation with either IQ score or scholastic success in later life.[22] Furthermore, childhood is known to be a time during which many brain and behavioural developments occur. These changes show a developmental time course and are reflected in the ability to perform abstract tasks and conceptualisations. Now even these tasks are the product of a similar ethos to that of the IQ testing — the presentation of what appear to be problem situations to the developing child to test his or her ability to cope with them — but nevertheless they do demonstrate that many of the concepts needed for the performance of IQ-type manipulation are learned at a time in the child's life when IQ scores are far from fixed.

It is astonishing despite all this that some educational psychologists continue to regard the tests, not as being one way of assessing an individual in a given social context, and in relationship to particular expectations and socially approved behaviour patterns — a possibly useful clinical tool — but instead as telling one something biologically real; the obsession with *g* is at best an index of the belief, held by many reductionist ideologues that if something can be mathematically expressed and manipulated, then it is therefore 'scientific' — part of the biologistic belief that a statistical phenomena implies a genetic mechanism. The 'general intelligence factor', is a property which emerges from multifactorial statistical analysis, and what Jensen, Eysenck and others do is merely reify a statistic. There is an old computer scientist's saying which is relevant here: 'garbage in; garbage out'.

Because this apparently quantitative approach to the question of intelligence essentially ignores what neurobiologists have been doing in the last decades in the field of learning, all the multitude of factors known to be involved in behavioural performance are discounted. To suggest that all of these may be combined to produce a single 'general intelligence factor' is as lacking in biological rationale as to talk of 'high IQ genes'. To go on from there to try to locate the 'habitation' of intelligence in the brain is, in terms of a scientific research programme, comparable to astrology or scientology. However, its ideological value gives it much greater social resonance than these follies.

GENETICS, DEVELOPMENT AND THE ENVIRONMENT

The relationship between the *genotype*, the individual organism's inherited complement of genes (DNA) and the *phenotype*, the expression of the genes in the actual organism itself, is complex and generally over-simplified, by biologism, which claims that an individual's genotype determines his or her *potential*, which is *modified* by environmental factors. Such a belief is a gross over-simplification of the actual situation because the definition of both phenotype and environment depends on the level of the analysis used — 'phenotype' may be used to mean an enzyme in one context, or a complex piece of behaviour dependent on myriad interactions in another. Such naive models reduce essentially to that of an empty vessel of a fixed size being filled to varying degrees during development, and of course it follows that once topped up, further filling is useless.

An organism begins at conception, at the combination of the genetic material from both parents and a small amount of egg nutrients. The genetic material consists of pairs of each of several thousands of genes, each of which carries the information to enable the cell to produce a particular protein. However, there may be several slightly different kinds of any given gene ('alleles') produced by mutations, and every individual is unique in its particular combination of genes. The odds against identicality — except for monozygotic (identical) twins who are both derived from a single fertilised egg — are astronomic. As the fertilised egg cell divides into many other cells, their pattern of growth is influenced by the environment in which they develop (for example the womb for humans or pond-water for frogs). Neighbouring cells produce chemicals which influence one another. Thus both the internal and external environment of the cell affects its genes. The result of this is specialisation: cells which were equivalent become different; they become brain cells, liver cells or skin cells. All the cells of an organism are each other's environment; they influence and are influenced, while the amount and type of nutrients available influence them all. In addition, the organism as a whole has an environment. But this environment is not fixed; animals seek food, heat or visual stimulation — they modify their environment, and it in turn affects them. The process is a continuous, never-ending dialectic.

At a time of Galton and Pearson, a gene was an inferred, abstract concept; today, however, it has chemical meaning — each gene makes a single protein, which has specific tasks to perform within the economy of the cell. What is the relationship between such a gene and a 'character' at the level of the organism? Sometimes it is relatively simple; eye or hair colour or blood group are determined by one or a few proteins. But how about behavioural characters like temperament or intelligence? Here the problem is much more complex because behavioural characters are measurements or abstractions from the properties of adapting organisms at a completely different level of organisation from that of gene or protein. Their biological correlates must involve the interactions of tens of millions of cells, hundreds of thousands of different proteins. The dichotomous approach of contrasting genetic and environmental causes is biologically naive because it fails to take into account the obvious reality that we have, at any point in time, an organism reacting with its environment — not just a gene. Moreover, the developmental dimension cannot be ignored, for it is in this context that we must assess the relationship between any gene or group of genes to the organism and its external environment. This means we have to consider not only the response of genes to their environment but the contribution of early genetic events to the 'environment' in which later genetic events occur.

Such a view does not ignore genetic variability; rather it embraces it. Certainly every person receives a unique constellation of genes at conception. But from that point onwards relatively few genes are active at any one time. The programme of switching genes on and off as development proceeds exhibits some paradoxical properties. Not only can the *same* group of genes in *different* environments result in different characters, but *different* groups of genes can, during development, result in the *same* character or structure being produced. This is because during development the biological machinery can adapt to certain contingencies and correct for them. However, overlaid on this capacity for self-regulation is the fact that there are critical periods throughout life during which the way in which the genetic programme responds to circumstances influences the patterns of response possible later in development. Examples include the fact that sex changes can be produced in the adult mammal, subverting its 'genetic programme', by minute temporary changes in the hormone balance in the young animal during a critical period. Indeed, for a certain type of mosquito a sex change can be produced merely by raising the temperature at which it is reared! The normal range of variation seen in human populations is an outcome of the way unique genetic groupings develop in their unique environments. Adaptability is a direct outcome of the interactions of genetic and environmental variability, and it is because this is both stable in some senses but subject to limitations by environmental contingencies that we say biological mechanisms are dialectical rather than digital.

Thus structures or activities are not 'in the genes'. This is why the proposition advanced by Eysenck[23] that the Black population in the

United States may be stupider than the Black population of Africa because when the slavers came over Blacks with 'high IQ genes' were able to escape, whilst the 'low IQ' Blacks were caught, is biologically as well as sociologically fatuous. Incidentally, it is noteworthy that, when he claims elsewhere that in Ireland the Irish population scores lower on IQ tests than does the English population in England, Eysenck deploys the reverse argument — the high IQ Irish were the ones who emigrated to America, leaving the irredeemably benighted bog-peasants behind![24]

There is no such thing as a 'low IQ' or a 'high IQ' gene — at best there may be particular combinations of genes which, in particular environments, produce 'high' or 'low' IQ. To take a simple example, there is a genetic disorder in humans known as 'phenylketonuria', a disease in which a particular substance, an amino acid, is utilised abnormally by the body. Phenylketonuric children are generally mentally defective. Hence the gene for phenylketonuria is a 'low IQ gene'. But if the phenylketonuric child is placed from birth on a diet in which the particular offending amino acid is absent, the child develops normally — in the different environment, the phenylketonuria gene is no longer a 'low IQ gene'![25]

This dialectical pattern of behavioural interaction is a normal part of the development of all animals. Parents and off-spring elicit and advance responses, and these are mutually modifying. Mice which have had their ear shape altered are treated differently to their siblings and, as a consequence, respond to their parents in a different manner.[26] Rats actively seek stimulating situations; they explore, and the results of deprivation or stimulation can be seen in various brain characteristics such as the amounts of important enzymes, or the degree of connectivity between brain cells.[27] The results of their earlier treatments can be seen in later behaviour; and how much richer than this is the human situation with its web of interactions and immense cultural variability? Within a given environment, an individual's behavioural patterns, once established, are fairly stable. The form that this stable structure takes is usually the result of a 'choice' at some critical devlopmental period, when the pattern to be stabilised was determined.

### HERITABILITY, GENETICS AND POPULATION DIFFERENCES

When pressed, adherents to the Eysenck/Jensen line would probably concede that, for any individual, the question of the contribution of 'genetics' and 'environment' to a particular trait like intelligence is not meaningful. Instead, they fall back to the position that what cannot be measured for a particular individual can none the less be measured for a population, so that in a given population the contribution of genes and environment to the difference (called the 'variance') between individuals can be estimated. The variance describes the variability of the trait; the more alike the population, the lower the variance.

To deal with the complex interactions due to many genes contributing to a character, the total variability is often crudely parcelled into two components, one readily attributable to environmental causes, and the remainder not. The ratio of the apparently non-environmental variance (genetic variance) to the total variance is called the 'heritability'. Heritability is a shorthand statement about the way the set of genotypes in a defined population tend to react in a defined environment. However, a fundamental question mark hangs over the whole contention that these estimates, which are essentially algebraic figures derived to interpret the results of breeding experiments for such quantities as yield of butterfat in cows or grain from crops, are applicable to attributes such as human behaviour at all. As a basic starting point for biometrical genetics, it must be possible to provide a phenotypic measure which, although it may show continuous variation, is quantifiable. Height shows continuous variation and a normal distribution, and is readily quantified against an essentially absolute scale, a ruler. Other characters such as blood grouping show discontinuous variation, but again are susceptible to absolute measurement. But IQ, even were it biologically based and not a social construct, would still not be a measure like height, or blood group; even its proponents only claim that it can be measured as a ratio of test score to population or age-group average, and it is hard to argue that an IQ is a phenotype in the sense that the biometric tools were designed for. Merely to pick an equation out of a textbook and apply it is no guarantee that the result will be meaningful, whatever the mathematical sophistication. What is more, it is often ignored, when heritability estimates are quoted, that at best they tell us something about the expression of gene or genes in a particular environment. Because it is merely an algebraic term, changing the environment may well change the heritability estimate. These estimates are figures which may be useful for animal or plant breeders working in relatively fixed environments; they are irrelevant in the context of the human situation, where as the environment changes, so too will the heritability estimate.

None the less, it is worth looking at how the famous 80/20 estimate for genetic versus environmental components in IQ variance are arrived at. The heritability of a behavioural trait can, it is argued, be estimated from studies of individuals more or less closely genetically related. Thus identical (monozygotic) twins have an identical genotype, while non-identical (dizygotic) twins do not. Much of the data on the heritability of IQ is based on studies of the rather few cases of monozygotic twins reared apart, when the environment can be assumed to be different whilst the genotype remains fixed. It is primarily from this type of study that the heritable contribution to the variance in IQ in White populations has been calculated as 80 per cent. Even were this figure valid in the limited sense we have indicated, it would be irrelevant for all practical or theoretical purposes. But is it even valid? The more the results on which it is based are re-examined, the shakier does it appear. Thus most of the original 'identical twins reared apart' data derives from a tiny number of studies made over many years by Burt and by Shields in Britain. However, a re-evaluation of these results has made it necessary to discard them on a variety of grounds (some of the 'reared apart' twins turned out to be brought up by an aunt in adjacent houses, in the same village and going to the same school), while Burt's data seem to have been 'adjusted' so many times as to be little better than totally false, as even Jensen now accepts.[28] The consequence is, to quote the conclusion of the exhaustive review by Kamin, that 'a critical review of the literature produces no evidence which would convince a reasonably prudent man to reject the hypothesis that intelligence test scores have zero heritability'.[29]

The extraordinary thing is, however, how many of the broader generalisations in this area are seen, when reference is made to the original sources, to rest on relatively little primary evidence. The point has been made by Kamin in the context of the Burt work; it has been made by Hudson[30] and by Hirsch — who accuses Jensen of almost incredible sloppiness and selective quotation in deriving his conclusions.[31] Most recently it has been made by McGonigle and McPhilemy[32] in examining Eysenck's claim that selection had operated so as to draw the genetically bright Irish to emigrate, leaving a genetically low-IQ population behind. McGonigle and McPhilemy, by going back to Eysenck's source material, a PhD thesis reporting the consequences of bilingualism on children's performance, show a consistent pattern of partial quotation, omission of crucial pieces of the original data and a 'failing to distinguish fact from speculation'; and conclude by charging Eysenck with 'professional irresponsibility in representing unqualified assertions as "facts" '. Thus the basic data on which the heritability estimates are made, far from being rock-like in solidity, are a morass of uncertainty.

But there is an even more fundamental flaw in the Eysenck/Jensen position than this. Heritability estimates are intended to measure the genetic contribution to the variance *within* a population – a biologically defined population with a freely intermixing gene pool. Measurement of how much the variance of a character is inherited *within* a population says nothing about the measure for the difference in that character between populations. Between two populations, the concept of the 'heritability' of their difference is meaningless. The genetic basis of the difference between two populations bears no logical or empirical relation to the heritability within populations and cannot be inferred from it. There is an example given by Lewontin which makes this clear. Suppose, Lewontin says, one takes two batches from a sack containing seed from an open pollinated variety of corn with plenty of genetic variation in it, grows it in pots containing vermiculite, watered with a carefully made up nutrient, Knop's solution, used by plant physiologists for controlled growth experiments; one batch of seed will be grown on complete Knop's solution, but the other will have the concentration of nitrates cut in half, and, in addition, we will leave out the minute trace of zinc salt that is part of the necessary trace elements (30 parts per billion). After several weeks we will measure the plants. Now we will find variation within seed lots which is entirely genetical, since no environmental variation within lots was allowed. Thus heritability will be 100 per cent. However, there will be a radical difference between seed lots which is ascribed entirely to the difference in nutrient levels. Thus we have a case where heritability within populations is complete, yet the difference between populations is entirely environmental!

But (continues Lewontin) let us carry our experiment to the end. Suppose we do not know about the difference in the nutrient solutions because it was really the carelessness of our assistant that was involved. We call in a friend who is a very careful chemist and ask him to look into the matter for us. He analyses the nutrient solutions and discovers the obvious – only half as much nitrates in the case of the stunted plants. So we add the missing nitrates and do the experiment again. This time our second batch of plants will grow a little larger but not much, and we will conclude that the difference between the lots is genetic since equalizing the large difference in nitrate level had so little effect. But, of course, we would be wrong, for it is the missing trace of zinc that is the real culprit. Finally, it should be pointed out that it took many years before the importance of minute trace elements in plant physiology was worked out, because ordinary laboratory glassware will leach out enough of many trace elements to let plants grow normally. Should educational psychologists study plant physiology?[33]

It follows that a prerequisite for any genetic study of the heritability of a trait between Blacks and Whites would be that the two groups formed a homogeneous biological population — that they more or less randomly intermarried and brought up their children in a colour-blind society.[34]

## RACE AND REDUCTIONISM

Thus one may conclude that, just as IQ measures can tell us little if anything about underlying biological mechanisms, so the apparatus of quantitative population genetics is a tool which cannot usefully be applied to an understanding of what contributes to the differences between individuals' performance on IQ tests. This is not of course to say that biology has nothing to contribute to the study of human learning or performance, or to adopt the straw-man 'environmentalist' position which Eysenck and Jensen often attack. Both genetics and developmental neurobiology have contributions to make to our understanding of humans — but not if attempts are made to use inappropriate tools to ask fundamentally unanswerable questions. There is, to put it plainly, *no type of scientific research programme which could be devised to answer the question 'how much do genes and how much does environment contribute to differences in IQ scores between different individuals or groups?'*. The question is not merely fallacious; it is, although *apparently* scientific, strictly meaningless; and no amount of leaving Eysenck in peace 'to advance research in the field of genetics', as the *Guardian* leader put it, would ever generate an answer. And, indeed, one may wonder why, under these circumstances, it has gone on being asked, from Galton to the present day, so obsessively but with such little scientific, as opposed to ideological, utility. To understand this, one must first look at the way in which the term 'race' is used in these discussions, and then look at the past history of the interrelation of racist ideas and intelligence testing.

The theme of race and biological reductionism runs throughout the Eysenck and Jensen work. It leads them to maintain that if there is a biological and a social phenomenon which are correlated, then the former *causes* the latter—it is like saying that shortsightedness *causes* studiousness.* In the lecture he was prevented from delivering at the London School of Economics, Eysenck, as has been pointed out, speaks of seeking for the 'habitation' of intelligence within the brain rather in the manner of a nineteenth-century phrenologist, and a similar 'causal' chain of thought can be found in such statemens as this of Jensen:

> The possibility of a biochemical connection between skin pigmenta-
> tion and intelligence is not totally unlikely in view of the biochemical
> relation between melanins, which are responsible for pigmentation,
> and some of the neural transmitter substances in the brain. The skin
> and the cerebral cortex both arise from the ectoderm in the develop-
> ment of the embryo and share some of the same biochemical
> processes.[35]

Despite the cautionary tone in which this is worded, these sentences are either quite devoid of meaning, or they are making the extraordinary claim that can be paraphrased as 'black skins may cause black brains'.

But the most damaging form of this reductionism occurs in the use of the concept of 'race'. There are two uses of the word 'race', one biological, the other social. The biological use of race refers to the relative discontinuities that occur in the distribution of a character within a species (for example, *Homo sapiens*); usually these different groups come to exist by evolving in geographically separate areas; the process does not produce absolute differences, but a gene pool, separated from other such gene pools; as a result of this separation. there evolve differences in gene frequencies between one race and another, though this need not be reflected in readily observable physical differences. Indeed, such variation *between* races is much smaller than variation *within* races. As Lewontin points out, 'less than 15% of all human genetic diversity is accounted for by differences between human groups! Moreover, the differences between populations within a race accounts for an additional 8·3%, so that only 6·3% is accounted for by racial classification.'[36]

By contrast, the social definition of race depends on social ascription, based on real or presumed cultural or physical differences: these differ from society to society and time to time; for example, in the 1930s there was a vogue (now discredited) amongst eugenicists for a division of Europeans into three 'races', so-called 'Nordic', 'Alpine', and 'Medi-

---

*And indeed it should come as no real surprise that a research paper has recently appeared claiming a correlation between 'the myopia gene' and IQ!

terranean'; today in South Africa, Japanese are classified as 'honorary whites', whilst Chinese are 'coloureds'; both the social characteristics ascribed to Jews and the legal definition of what constitutes a Jew has differed between, say, Germany in the 1930s and Israel today.

What Eysenck and Jensen do is sometimes to use the biological and sometimes the social definitions with precision; at other times, however, the two meanings are interwoven and used with characteristically ideological ambiguity. Thus, when he wants to emphasise the biological respectability of his analysis, Jensen goes into great detail about population genetics within biologically defined races. But in going back to his raw data on educational performance, his categorisations are socially based. US Blacks (with their estimated 25 per cent of 'White' genes) are defined socially, not in biological terms, so that a Black is one who is classed by the society in which he lives as 'Black', just as in the apartheid situation of South Africa. At other times Jensen and Eysenck attempt to treat Mexican Americans, Irish, Jews — and even the working class as a whole — as if these represented races in the biological sense — the only sense in which it is possible to study population genetics. This biological reductionism is precisely, of course, that used by the Nazis and present-day fascist groups. The oscillation of Jensen's frames of reference between the biological and the social is typical of the biologism which dresses ideology as science and hence claims an unchallengeable basis in 'objective scientific facts'. It is precisely because, therefore, the questions which Eysenck and Jensen wish to ask cannot be given a 'scientific' answer that we must look more closely at their historical antecedents.

## THE INVISIBLE COLLEGE OF THE MENTAL TESTERS, 1869–1969

The history of mental testing,[37] and the pursuit of the most intellectually talented, began in the modern age with the work of Francis Galton. While Plato's *Republic* was a philosophic precursor, *Hereditary Genius* (1869) was the first of a series of studies on the inheritance of intelligence. Galton studied the relations of a variety of eminent men (1 in 4000 of the population of Victorian England, he estimated fell into this category) and showed conclusively that judges, statesmen and divines, literary men and scientists, tended to have among their relatives, often stretching back through several generations, other judges, statesmen, and so forth. Here, Galton concluded, was incontrovertible proof that genius was inherited, and of inherited genius, Britons above all, and other European races to a lesser degree, were disproportionately endowed with other races.

Galton was not merely the first to attempt to prove that genius was hereditary, that intelligence was genetically determined. His work predates the understanding of the biological basis of inheritance, which depends on the rediscovery of Mendel's work in the first years of this century, followed by the successive unfolding of the 'biological revolution' which culminated in the 1950s in the recognition that the 'double-helix' of DNA represented the genetic material. Galton was none the less able to devise an elaborate mathematical treatment to study the variability of particular 'characters' (such as height, hair colour, intelligence, and so on) within the population. So convinced was he that most human qualities were inherited that he and his pupil Karl Pearson founded a new subject — later dignified by the name of science — *eugenics* — devoted to attempts to propagate a 'healthier' race by using techniques comparable to those of the stockbreeder. In the name of eugenics, the middle class was to be encouraged to breed, to prevent being swamped by the working class, just as more than half a century later Vorster in South Africa urged the Whites to outbreed the Blacks.

The eugenicists applied themselves to protecting both the middle class from hereditary criminals, vice and political agitators, and the imperial countries (conceived of not merely as nation-states, but as biological races) from inferior races. For Karl Pearson, as we earlier noted, 'The path of progress is strewn with the wreck of nations; traces are everywhere to be seen of the head tombs of inferior races . . . the stepping stones on which mankind has risen to the higher intellectual and deeper emotional life of today.'[38]

As eugenics crossed to the United States, it became even more explicit. Of the US Blacks we find the eugenicist Holland claiming in 1883 that 'Galton's law is squarely across their path, and the sooner they die gently out the better, and to assist them to multiply becomes as wrong as keeping the filthy and effete Turk in Europe for the sake of containing Russia.'[39]

Thus the study of the origins of intelligence was firmly secured within a Social Darwinist mould, and it is precisely this preoccupation with inter-group struggle, whether of race, nation or class grouping, linked to increasingly sophisticated statistical techniques, which was to become the hallmark of the invisible college of mental testers. This college was linked not only cognitively but socially too. It is not merely that Eysenck is the pupil of Burt, the pupil of Pearson, but the lineage crossed to the United States by way of James Cattell, father of statistically orientated American psychology, who studied first with Wundt in Leipzig, then became a co-worker with Galton in England, and shared a common belief in the hereditarian view of intelligence. The lines rejoin with Jensen, who in his turn has worked with Eysenck.

The early eugenicists talked of intelligence in very general terms of superiority and inferiority, as they had no means of quantifying it. The techniques for doing this were developed at the beginning of this century, notably by Binet in France and Burt in England. The quantitative measure that they developed, the IQ was, in Binet's hands, originally intended as a way of classifying school children to provide special educational help for children who performed poorly. Binet's work was adapted and introduced to the United States by Henry Goddard in 1908, extensively revised by Lewis Terman, and presented to the world in 1916 as the classic Stanford – Binet scale. However, although Binet's personal motivation seems to have been clinical, it had at its heart a biological theory of intelligence which made it particularly attractive to the psychologist hereditarians such as Goddard and Terman in the United States, or Burt in Britain. For Britain, the IQ tests served merely to 'confirm' the Galtonian hypothesis that the middle classes were superior, as, by and large, they scored higher than the working class; but they also showed the 'potential' of many working-class children to rise, for the overlap in scores between the two groups was considerable. As the biological inferiority of the Blacks was taken for granted, and, anyhow, they did not actually live here, it was at that point not necessary to direct the weapon of the I Q test towards them. The situation was very different in the United States, where the race issue was not separated by a strip of water and an imperial tradition, but was present, and increasingly so with the internal migration from South to North, in every state and city. There, IQ testing was taken up with enthusiasm; a generation of child and educational psychologists modified and refined the techniques – a generation whose leadership demonstrated an overt and widespread eugenicism.

Goddard, argued that there was a close correlation between social-class position and the level of intelligence, and went on to conclude that those at the top of the pyramid had a particular responsibility to care for the masses in terms of 'social welfare' and 'national efficiency', and that in the context of a democracy it was important that the intellectual elite persuaded the masses to submit to their leadership. For Goddard the virtue of the IQ test was that it demonstrated that he lived in a meritocracy, where all citizens stood in the estate to which biology had been pleased to call them.[40]

Nor was Terman's contribution ideologically at odds with his predecessors; his eugenicist concerns were supported by his belief that the social system was founded on the distribution of IQ:

> The racial stocks most prolific of gifted children are those from northern and western Europe and the Jewish. The least prolific are the Mediterranean races, the Mexicans and the Negroes. The fecundity of the family stocks from which our most gifted children come appears to be definitely on the wane. This is an example of the differential birthrate which is rapidly becoming evident in all civilized countries. It has been figured that if the present differential birthrate continues 1,000 Harvard graduates will, at the end of 200 years, have but 56 descendants, while in the same period 1,000 S. Italians will have multiplied to 100,000.[41]

Other Americans pre-eminent in the psychological testing movement, such as Yerkes, Hall, Brigham and Thorndike, were equally unequivocally racist in their conviction that the Blacks were genetically inferior to Whites. Nor, apart from their commitment to testing, were their social attitudes markedly different from mainstream U.S. psychology — William McDougall was, for example, a leader of the instinct school of psychology and a keen eugenicist, believing that he could demonstrate racial differences in intelligence on the basis of differences in skull size.

The IQ studies on American soldiers in the First World War seemed to support their views. The belief that Blacks had inherently inferior intellectual capacities and that they 'needed' a special industrial education is a time echo of Jensen's view of the Black's ease at associational learning and difficulty with conceptual learning. 'It was therefore possible', Pickens notes, 'for an individual to advocate themes of progressive education (education for the child's unmet needs and desires) and at the same time give scientific support for the racist legacy of the 19th century.'[42]

While the spokesmen and leadership of the eugenicist movement were not recruited directly from among the psychologists or from the geneticists, the view that the science of heredity would open the way to an improved humanity was shared by all. The eugenicist movement's heyday was from 1905 to 1950, much stronger, more overtly racist and pro-Nazi in the United States than in Britain, where it tended to be feudalistic and a good deal less effective. With the emergence of a powerful group of British human geneticists, notably around J.B.S. Haldane in the 1930s, the genetic inadequacy of the British eugenicists' position was made manifest. The ideological grip of the eugenicist argument was none the less powerful; Beveridge, the architect of the 'welfare state', was quite prepared to deny the right to parenthood to the lower strata on the grounds that the national intelligence was declining, and a variety of proposals for sterilising the unemployed — as well as the mentally defective and habitual criminals — were canvassed. However, in the United States, solutions to the 'menace of feeble mindedness' went well beyond the debating stage. In some thirty-one states, sterilisation laws were passed, and by 1935 20,000 sterilisations had been carried out under these laws.[43] Perhaps even more important than the compulsory sterilisation campaign was that mounted for selective immigration on grounds of racial inferiority and danger to the national stock through racial miscegenation. Under the leadership of Harry McLoughlin, (eventually to be given an honorary M.D. degree by the University of Heidelberg in 1936 — then a centre of Nazi race theory) the psychologists played their part with the geneticists in providing the scientific rationale for the selection policy.

It was in Germany, of course, with its Eugenic Sterilisation Act (1933) — the forerunner of the gas chambers — and the clear-cut ideology of racial superiority and 'Aryan biology' of the Nazis, that the previous history of eugenics and scientific racism ends. Racial biology and psychology were proclaimed by biologists (as in the quotation from Lenz given earlier[5], by theoreticians like Rosenberg, and even Nobel prize-winning physicists like Philip Lenard and Johannes Stark, who issued statements in the 1920s supporting claims of Jewish genetic inferiority (other inferior groups were of course, Slavs, gypsies and Blacks).

What is interesting is how, despite the distance in time and the advances in human genetics, we are now asked to return to the intellectual proccupations of the eugenicists and mental testers of a past age. When 100 years after the publication of *Hereditary Genius*, we are asked to reconsider the nature/nurture debate, it must surely be clear that we are not dealing with a scientific question at all, capable of scientific resolution; the issues which preoccupied Galton and Pearson, Thorndike and Terman, and which now re-emerge in the sophisticated dress of the 1970s, are derived not from scientific issues suspectible to resolution but are primarily reflective of political and social concerns.

It is thus to the socio-political framework of scientific racism which we must turn to set both the current and the past issues in perspective.

## THE SOCIAL AND POLITICAL FRAMEWORK

As one studies the present and past history of the issues raised by the race/IQ debate, it is impossible to avoid the links between its periodic emergence and changing external political circumstances. The role of Social Darwinism in providing a biological rationale for the structure of Victorian society is well known. Darwinian evolutionary theory drew on the ideas of Malthus about human population to describe competition within biological species, but also found its own biological concepts pressed into service as Social Darwinism, which took in distorted form such evolutionary slogans as 'the struggle for existence' and 'the survival of the fittest' and fitted them to *laissez-faire* Victorian capitalism.[44] The middle class, the fittest, had a biological as well as a social right to their privilege, and intervention in this process by, for example, the state, in educating or protecting the health of the working class, was defying biology — going against nature. The middle class began to see the working class almost as a separate race — shorter and less healthy, with alien language and culture — a race born to *be* workers.

What Galton provided was the beginnings of the method of quantifying the rights to such privilege, and, by way of eugenics, a possible strategy for retaining it. What has this to do with race? The relationship between the publication of Galton's work and the height of Victorian imperial expansion, with all its attendant ideology, is too close to be ignored. It was not difficult to see colonialism in pseudo-Darwinian categories; the peoples of other countries were different races, the biological imperative meant that races should struggle competitively to survive; the fittest won. As the English were the fittest, it was only natural that they should win, and their civilising mission abroad was thus supported by a biological imperative — the colonised races were inferior in virtually every respect. The advantage of racism was that it is indifferent to class. All the English — even the working class — were superior to, and more intelligent than, the 'lesser breeds without the law', the 'wogs' who began at Calais. Galton's work is unequivocal on this; it is full of such phrases as 'from the highest Caucasian to the lowest savage'. Here began the fatal conjunction whereby racism, by stressing one nation, the imperial country, against its colonies, has been able for so long to mystify and divide its working class to the continued

F

benefit of its ruling class. It is of crucial importance to remember, therefore, that from its very inception in the nineteenth century, the technique of categorising people according to their apparent biologically determined intelligence has been used as a method of justifying both the class structure and the racial discrimination of imperialist societies.

None the less, so far as the British situation was concerned, the major concern of eugenics and the IQ movement in the first part of the twentieth century remained one of class rather than race; thus when Burt introduced the Binet tests into the British situation, he did so as a meritocrat; believing, as he continued to do until his death, that most of intelligence was genetically determined and located in the middle class, none the less he could see that the distribution of intelligence must mean that there were a substantial number of working-class children whom the system was excluding; the tests were a way of screening the working class, on a meritocratic basis, so as to select out that proportion and offer them the chance of a grammar-school education, a particularly important role in Britain where a rigid educational structure had become dysfunctional in a society where there was an urgent need to expand the base of a technologically skilled labour-force.[45]

The situation in the United States was different; as even the brief history of the testing movement we have given shows, its development was from beginning to end linked to two explicit socio-political goals: the control of European immigration and of the ex-slave Black population. Its roots long predated the psychologists. 'Ethnology', or as Josiah Nott charmingly called it, 'niggerology',[46] had earlier justified slavery by referring to the Blacks' innate capacity for knee-bending, or the inherited disease of drapetomania (tendency to run away). The scientific racists of the 1920s and 1930s developed a tone of pious resignation, the regrettable acceptance of biological laws of human inequality by hard-headed realists with the self-appointed task of biological law enforcement.

None the less the 1940s saw the end of this period of scientific racism. The challenge of Nazism domestically and internationally resulted in a mobilisation not merely of the working class in Britain, but also of many scientists and other intellectuals, who saw that the cultural battle over genetics was part of the larger conflict with Nazism. The fallacies and ideological role of Nazi pseudo-science were energetically exposed, and the bulk of the British scientific community rejected men like Lenard and Stark with contempt. The conflict of the 1939–1945 war, which buried Hitler's thousand year Reich, submerged most of its racist biology as well, and the years of 'end of empire' which followed 1945, with the emergence of many new African or Asian nation-states, seemed to have completed the process. For a period at any rate, until the new relationships of economic neo-colonialism could be established, most of the capitalist West was on the defensive against the new nations; expressions of a supremacist type were apt to be bad business and were frowned upon. Many believed the ideological battles of the 1930s were over. The atmosphere of the period is well summed up in the conclusion of a massive UNESCO study which affirmed in 1951: 'According to present knowledge, there is no proof that the groups of mankind differ in their innate mental characteristics, whether in respect of intelligence or temperament. The scientific evidence indicates that the range of mental capacities in all ethnic groups is much the same.'[47]

It was this period of ideological truce which events of the 1960s were to shatter; a series of significant defeats for white imperialism, of which the Vietnam War, with its racist overtones, was but one; the sharpening conflict with neo-colonialism, old-fashioned colonialism and *apartheid* in Africa, the deepening race conflicts in the United States, and the quite new tensions created by the fact that in Western Europe the economic upsurge of the late 1950s and early 1960s relied in part upon an influx of cheap immigrant labour, Turks, Yugoslavs, Southern Italians, Spaniards, Algerians, and into Britain, West Indians, Indians and Pakistanis. As the economic situation deteriorated, so the social problems have sharpened, and racial tensions between immigrants (*gastarbeiter*) and the indigenous population have become exacerbated in many Western European countries. In Britain the black immigrant group and their children face ESN labelling in schools, and unemployment or restricted entry into skilled jobs in industry.

It is then possible to suggest some of the reasons why there has been a revised interest in eugenicist and racist theories of intelligence. It is not that these ideas were absent during the period from 1945 onwards; many educational psychologists, particularly in the United States, continued to 'prove' the inferiority of the Blacks — Audrey Shuey's voluminous work[48] is but one example. However, the question is why they resurfaced so publicly from 1969 onwards. It does not seem unreasonable to suggest that the new emergence of interest in old ideas is associated with similar social contradictions; indeed, Jensen's own (1969) paper is very specifically related to one such: the failure of the American Poverty Program, and in particular Project Head Start. At a time when American society has clearly failed to find a resolution to its

F*

internal conflicts, such as the decay of the inner city, with its increasing urban violence and underlying racial antagonisms, there has been a tendency to seek once again for biological explanations for social issues. Ethologists talk of innate aggressive behaviour, psychosurgeons propose amygdalectomies to 'cure' ghetto violence and Herrnstein's and Jensen's arguments tell us that social stratification follows IQ, and if Blacks (or working class) are underemployed — or unemployed — then the fault lies not in their stars, or in the social order, but in their genes.

## WHAT DOES SHAPE HUMAN PERFORMANCE?

We have shown that the whole edifice of 'genetically' based black/white or working-class/middle-class intelligence differences is no more than ideology, deriving from and serving to sustain the present social order. IQ is the symbol of a society which is determined to perpetuate class distinctions, where even the liberal hope of equality of educational opportunity is so far from reality as to raise no more than a hollow laugh. By and large, children's education in Britain has always been designed to fit them to their class destiny, with the bare minimum of mobility available to keep the pressure low and service technocratic needs. The disparities in the system are there in the primary schools, compounded in the streamed secondary schools or comprehensives with their appended depositories for the educationally subnormal[49] and capped by the superstructure of 'higher' and 'further' education. The structure is there with or without the IQ testers who descend upon it with their self-fulfilling prophesies to demonstrate its apparent biological inevitability.

But there are important — and scientifically valid — things which biologists *can* say about behaviour and even intelligence, which are studiously ignored by Eysenck and Jensen, intent on grubbing through the discarded eugenics of the 1930s. There is a wealth of neurobiological evidence which reveals the way in which environmental factors, particularly during infancy, can affect not merely behaviour, but also the very structure of the brain itself. Such factors may be relatively crude, like malnutrition, or much more subtle, like the quality of the environment.[50]

What this means is that amongst the predictive factors for a child's subsequent performance at school are included family size, socio-economic status, the mother's health during pregnancy and child's birth weight. Put more directly, the best way of ensuring that a child has a low IQ is to raise him or her in poverty, with inadequate food, poor health and bad home conditions. Where these conditions are improved, then school performance and 'IQ' improve — as in the well-known comparison of two generations of 11-year old Aberdeen school children, on tests between 1932 and 1947.[51] Whereas the Eysenck/Shockley type model, that the national intelligence is declining, would have predicted a fall in IQ between the two generations, the actual results show a small increase.

Even the performance of individual children can be dramatically shifted by changing the environment. In an often-quoted study by Skodak and Skeels,[52] the IQs of a group of white children put into 'high socioeconomic status' (SES) adoptive homes, mostly before they were six-months old, were compared to those of their (low IQ low SES) biological mothers. At age thirteen or fourteen the children showed an elevation of fully twenty points, from 85·5 average to 106 average. In another example, cited by Eysenck, Heber studied poor Black children with an expected IQ of 80, placed in a special school with a great deal of personal interaction with a trained social worker; the children's IQs, in Eysenck's own words, 'simply shot ahead' to 'well above the 100 mark'.[53]

The point is not that biological factors *cause* good or bad performance on IQ tests or other sorts of behavioural measure, or that intelligence is *independent* of a biological base, but that present-day neurobiology is coming to understand that there is a continual interaction between the biology of the individual and his or her environment. Bad schools and an alienated society have biological as well as social consequences.

## CONCLUSIONS

We have tried to show that the scientific basis of Eysenck and Jensen's position is not what they claim. Their 'evidence' says nothing about the question of genetic differences between populations, although this is the cornerstone, the *raison d'être*, of their analysis. We are faced with a series of questions about human behavioural propensities; questions which raise issues about the human brain and its development. This is something we can say things about. In a sense the brain can be considered an ideological organ — it not only stores and enables the cultural transmission of our ideologies, but its development and performance can be shown to be a function of the social context in which the individual develops.

This is true not only in the sense that the correlates of the unequal distribution and control of resources (for example malnutrition) can include deficits in growth and performance, but also in the sense that the perceptual and cognitive performance of the brain will both respond to the world view of the society in which the individual grows up, and contribute to this world view. Inevitably the critique of Jensen and Eysenck must move on from an attack on their biological determinism to the impoverished world view that underlies, promotes and is the reason for their analysis.

If we are really — as Jensen claims he is — interested in 'boosting IQ and scholastic achievement', and all humans are the product of an interaction of their genotype with their environment, we can in principle do one of two things: modify genotype or modify environment. No one knows biologically how to modify genotypes, or in what direction to do so, and the only sort of society in which such modification could be achieved would be a Nazi one — or a Marcusan situation of repressive tolerance coupled with cash incentives. On the other hand, we do know how to modify environments. We could do quite straightforward things, like eliminate malnutrition, poverty, slum schools and the self-fulfilling prophesies of teacher labelling of children for a start. We could go on to eliminate the environment which attempts to reduce not merely children but adults from thinking, creative humans to alienated 'hands' divorced from all but the most routinised thought — the life experience of the vast majority of the population of Britain or the United States. The logical consequence of being in a society whose mode of production demands cultural and material alienation is precisely the diminution of the creative potential of the great majority of its members. Thus a prerequisite to answering Arthur Jensen's question, even if we do not regard *his* sort of scholastic achievement as our goal, is the transformation of the society in which we live. It is because of this that, at times such as the present, of acute internal contradictions within capitalism, issues of race and IQ come to perform a twofold ideological role.

One the one hand, they provide an apparent 'scientific' rationale for the existing social order. If we live in an hierarchical, alienated society, a society in which some are superior, others subordinate, this is portrayed as conforming to a 'biological imperative'. The distribution of the IQ score conveniently parallels the social order. The ideological role of this biological imperative is as manifest — though more sophisticated — as was the use of evolutionary theory in the nineteenth century.

At the same time, the race/IQ issue performs another role in the service of capitalism — it is by its nature divisive; it sharpens not merely the class division of society, but also, within the working class, helps exploit the division between Black and White. Where what is needed is the unity of the class in its common struggle, scientific racism, manipulated and exploited by the media, helps foster prejudice and tension within the class, sustaining racism even within the trade unions and the labour movement.

But just as human beings are not genetically programmed machines, or aggressively overcrowded rats, they are not condemned to exist in a class and race-bound capitalist order into the interminable future. What distinguishes humans from other animals is their social existence and their capacity under certain conditions to transform their own society and hence their own mode of existence. Rigidly defined categories such as IQ do not allow for the transformations of human consciousness which are produced in struggle and revolution — those creative turnings upside down of humanity's social environment. In the act of mobilising against the ideological use of pseudo-science in human oppression, we can help both to liberate ourselves and to transform society.

# 8
# Women's Liberation: Reproduction and the Technological Fix

## Hilary Rose and Jalna Hanmer

We wrote this chapter for three very simple reasons: that human reproduction is necessary for the continuance of the species; that a significant section of the women's movement sees reproduction as the source of female oppression; and that science and technology are continuing to make rapid advances in the regulation and modification of reproduction. Our particular concern is whether these advances, given that science and technology are part of the culture and institutions of a class-bound and patriarchal society, are likely to add to women's liberation or to their oppression.

Advances over the past one hundred years in the science and technology of regulating reproduction have created the possibility that women, without limiting their sexual activity, can choose if, and when, to bear children. This 'choice' is at the level of technical feasibility, choice in actuality is limited by religion, class, race, husbands, knowledge, state action or inaction – to name but a few key factors. However, science and technology do not stand still, advances in knowledge continue to be made at an ever-increasing rate, and, most importantly for the purposes of this chapter, the gap between new knowledge and its implementation is shortening.

Yet despite the relevance to women of much of the new work in, for example, developmental biology, science and technology have been either little or relatively uncritically discussed within the women's movement. To a considerable extent the exclusion of science and technology from the movement's agenda is explained by the exclusion of women themselves from the hard (that is masculine) sciences. Apart therefore from a certain amount of descriptive work showing how women are hindered from entering the institutions of science, and are at a status and financial disadvantage even when they manage to gain entry,[1] the movement has not extended its analysis of the male-dominated culture to consider the culture of science. In what follows we critically analyse those strands within the movement which discuss reproduction, point to the implications in this area of the theoretically unresolved question of the relationship between the women's struggle and the class struggle, and set the whole in the context of the science and technology of regulating reproduction both presently being developed and also being planned.

## WOMEN'S LIBERATION ON REPRODUCTION

Within the women's movement are two major theoretical wings: the radical feminists and the feminist Marxists; the former see sex as the primary contradiction, the latter some combination of sex and class. Each wing is faced with its own central theoretical problem. For the radical feminists, who advocate political lesbianism, the central problem is that of the continued reproduction of the species. For the feminist Marxists, the problem is to articulate a necessary relationship between the liberation of the class and the liberation of women. Consequently within radical feminism there is occasionally expressed a belief that science and technology can be used positively to reorganise and eventually eliminate natural reproduction; whereas for the most part the feminist Marxists accept natural reproduction, but seek to alter the social forms and ideology surrounding it.

### Radical Feminism and the Technological Fix

One main thesis of radical feminism asserts the biological basis of women's social inferiority, and has been most forcefully developed by Shulamith Firestone. Her book, the *Dialectic of Sex*, has precipitated an important debate on artificial reproduction both within, and without, the movement. Her view is that it is because women bear children that it has been possible for men to gain ascendancy over them, for the subjugation of women is rooted in the division of labour which begins with the differing roles males and females have in the reproduction of the species.[2] This division of labour is institutionalised in the family.[3]

Therefore, to free women it is necessary to eradicate the family first through developing alternative life styles and social institutions and eventually by reproducing people artificially, eliminating the female reproductive function. Equality for women is to be accomplished through scientific discoveries that progress from the artificial reproduction of babies to the elimination of childhood, ageing, and eventually death itself. The technological fix ensures the new Utopia.

Firestone argues that certain current social, economic and technological trends support the changes she is advocating. The family is under a death sentence from two sources. First, there is no longer a need for universal reproduction, and, second, the coming cybernetic mode of production will make the family as a unit of reproduction and production obsolete. Cybernetics alters human beings' relations to work and even the need to work, and this change in production will eventually 'strip the division of labour at the root of the family of any remaining practical value'. Thus the woman will no longer be needed to service the paid male worker (and his children) as in present-day assembly-line technology.

Firestone points out that there is no image of the perfect society created by feminist women, nor even any Utopian feminist literature. She suggests that the perfect society for women would involve flexibility, multiple role options that exist simultaneously and also can be chosen serially. One option during the transitional phase towards the ideal society would be the opening up of professions that satisfy the individual's social and emotional needs, so that the pressure on women to establish family units is not so great. Another transitional solution is to encourage 'deviant life styles', which imply non-fertility, such as relationships between two or more people, of the same or different sex, living together in groups. She thinks that after several generations of non-family living 'our psychosexual structures may become so radically altered that the monogamous couple . . . would become obsolescent'.

To some extent her solutions are shared by other theorists of the family as diverse in ideological orientation as Mitchell[4] and Packard.[5] It is also the case, as reported by Skolnick and Skolnick,[6] and Abrams and McCulloch[7] that people *are* living in more diverse household or 'family' structures. As we shall argue later concerning various reproduction technologies, the liberatory or repressive character of these forms is ambiguous, to be determined not only by those who hold the power but also by the ideology embedded in and surrounding them.

*Marxism and Women*

Firestone's thesis stressing the primacy of the sexual division thus stands in direct opposition to the Marxist tradition, which (with the possible exception of the Chinese revolution) has followed the Bolsheviks and stresses the primacy of class. For Marxist women in contemporary Britain the choice is sharply posed. Either they can join one of the various male-dominated Marxist groups or, eschewing this dominance, they can become active within the women's movement but as non-aligned Marxists. By male dominance we mean more than the obvious aspects, such as exclusively or nearly exclusively male leadership (particularly as theoreticians, while the women's contribution is limited to administrative tasks); we mean also the male ideological hegemony. Consequently the act of choosing to join appears to place the interest of the male industrial worker — which is seen as synonymous with 'the class' — above and separate from the interest of women. The alternative, of remaining non-aligned but active in the women's movement, is in effect to work in a movement which by its class composition is to a great extent divorced from working-class experience and struggle.

Orthodox Marxism thus sees the class struggle as primary, a concern for the position of women as secondary: an issue which will be resolved — and, depending on the faction, possibly only thought about — after the revolution. This is no recent development; it arose prior to the degeneration in Marxist thinking associated with Stalin (although with his medals for fecund Stakhanovite mothers Stalin played his part). It is, for example, expressed quite unambiguously in Lenin's exchanges with Zetkin and Kollontai during the early years of the revolution. Whether it was the practical issue of famine and civil war or a more profoundly embedded view of Lenin's is not material to the argument. Lenin clearly put to one side the ideas of free love and the withering away of the family as expressed in the *Communist Manifesto* as a bourgeois preoccupation. Instead, like Engels, he saw the solution in terms of women entering the productive process as workers; liberation would, in some unspecified way, flow from this. The nature of this 'solution' is illustrated by the present situation whereby women in the Soviet Union, despite some necessary assistance with child care in nurseries and creches, merely carry out double work — in the work-place *and* in the home.[8]

Whatever other disagreements developed between Stalin and Trotsky, both were the heirs (note the sex) of Engels and Lenin in their adherence to this economist solution, and neither the orthodox communist parties in or out of the Soviet Union nor the various Trotskyite groupings differ sharply from this view.

The theory and experience of the Chinese revolution was different. At a theoretical level the question of sexual (and generational) repression within the family and for the need for the struggle to be anti-patriarchal is discussed within Mao's early writings. In addition, within the history of the revolution itself, the part played by the profoundly oppressed peasant women was crucial, as documented for example in Jack Belden's *China Shakes the World*,[9] or to a lesser extent Hinton's *Fanshen*.[10] Against the prostitution and slavery of women in both Imperial and Chiang Kai Chek's China, the puritanical monogamy of post-revolutionary China represents a major advance; however, there is little diversity tolerated in either sexual expression or social forms. Because women are, in theory, 'half the sky', the Chinese Communist Party does not see the struggle against sexism as a class struggle. In practice they are reticent in articulating their analysis of the nature of the issue; sexism seems to be seen as a contradiction among the people rather than between the people.[11] This, as we shall discuss later, has had profound implications for the planning of reproduction.

With the exception of the Maoists the revolutionary left takes little theoretical interest in this discussion. However, while in Europe many countries have experienced considerable growth in Maoist groups and movements, in Britain they have remained trapped in a multitude of groupuscules, each too small to make a significant contribution to this debate. The challenge of the radical feminists who argue that women *are* the class is thus in practice for the most part ignored. Their analysis is felt to be of such little theoretical interest that it is only worth the attention of women's sections, certainly within the bigger Marxist groups.[12] Consequently, for any systematic discussion of Firestone's and other radical feminists' theses and their relationship to a class analysis, it is necessary to consider the contribution of the feminist Marxists.

*Feminist Marxists*

For the feminist Marxists there is considerable difficulty in linking the two concepts of 'sex' and 'class'; either they are opposed or they operate on different planes. However, despite the problem of specifying the relationship between feminism and the class struggle, for the women Marxists the attempt, and belief in its success, in an essential credo:

Within the women's movement therefore, we reject both class struggle as subordinate to feminism and feminism as subordinate to class struggle. Class struggle and feminism are for us one and the same thing, feminism expressing the rebellion of that section of the class without whom the class struggle cannot be generalised, broadened and deepened.[13]

The position of the feminist Marxists is in this respect not totally different from that which Lenin found himself in. Far from indifferent to the question of women's liberation, he was unable to articulate the relationship between the liberation of the oppressed sex and that of the oppressed class. Thus, like Lenin, the feminist Marxists all too often have to invoke an invisible hand of socialism which will ensure that through the revolution both the women and the class are liberated.

Juliet Mitchell's work, which draws extensively on Althusser, points to a possible method of analysis and action.[14] Instead of discussing the woman's condition as a monolithic entity, Mitchell seeks to specify the separate structures which together form the complex whole. Each separate sector is to some considerable extent autonomous, and each has its own momentum; therefore the complexity created by the synthesis of all the structures is itself continuously shifting. Mitchell then distinguishes the four basic structures as: 'production', 'reproduction', 'sex' and 'socialisation'; each is autonomous and therefore requires discrete analysis, yet at the same time all four are linked in the complex continuously changing totality which is the woman's situation. Sometimes the movements in the separate structures cancel each other out, but at the moment when the separate structures reinforce one another and intensify the contradictions, the conditions for revolutionary change exist.

Selma James' and Maria Rosa Dalla Costa's critique[15] of the inadequacies of past and existing Marxist theory and practice is particularly sharp. Dalla Costa, for example, points out that Gramsci denies that working-class women are part of the class, and speaks of the need to 'neutralise' them, as he sees the male industrial workers as *the* class. Dalla Costa rejects the view that women are not engaged in social production; housework, including child-bearing, creates labour power and thereby value. It does not seem to be so only because women do not receive a wage and are trapped in pre-capitalist forms of production. What is peculiar to women's work, as countless feminists have observed, is the social isolation which stems from domestic labour. However, to compress the separate structures of reproduction and production has theoretical disadvantages. Where women are increasingly engaged in production (in the conventional sense) and decreasingly engaged in reproduction, the need for a conceptual distinction between the two activities increases rather than decreases.


## Radical Feminism's Consort to Conceive

Carla Lonzi of Rivolta Feminile[16] writes as a radical feminist, but whose theoretical stance is profoundly influenced by a Marxist analysis — quite unlike that of Firestone, who regards science and technology as autonomous. Like Mitchell she separates reproduction from sexuality in an analytical sense, but unlike Mitchell does not regard the split between them as anathema to bourgeois ideology. Perhaps this is because Mitchell's original article was pioneering work published in 1966, whereas Lonzi's *We Spit on Hegel* was published in 1970, when both the situation and women's consciousness of the situation were different.

Rivolta Feminile's documents (some are collective and anonymous) argue that a male world is able to turn this separation to a male and bourgeois advantage, and speaks of men 'colonizing women through the penis culture'. Despite all the talk about age not mattering in heterosexual relationships in practice it only 'doesn't matter' one way. The advantage men gain from feeling and being able to draw on a potential pool of all women younger than themselves (decrease in male physical attractiveness being compensated for by high status, money, and so on), while most women feel, and are, unable to similarly consort with men twenty or thirty years younger than themselves, loads the dice against women. In affairs the age difference and therefore the power element is often even more marked than in marriage, so that the split between sexuality and reproduction made possible by contraception and abortion tends at present both to aid women but also to increase male hegemony. Men obtain sexual satisfaction, their dominant status being amplified by permissiveness — and, because affairs are predominantly infertile, the population is limited.

Rivolta Feminile's recognition of this is expressed in their thesis that women should consort with men sexually only for the purpose of reproduction. Thus, in the interests of the liberation of women, assessing the present relationships of the structures of reproduction and sexuality, they argue a position on heterosexual contact remarkably like that of St Paul. Real sexuality and tenderness is seen as something to be shared only with women; men's utility is to be limited to the act of fertilisation. Biological necessity is recognised but the social forms are to be completely refashioned.

The question of reproduction thus is set constantly against the position of women and their struggle for liberation. Within the movement, whatever the tendency, the need for women to be able to control their own bodies (that is the reproductive function) is seen as a primary and crucial step towards their freedom. They see discussions of the control of population, whether to increase, to stabilise or to decrease it, as a societal abstraction, managed and dominated by male others.

Thus, while the women's movement all over the world seeks in some general sense to increase women's control over their own bodies as part of their wish to control their own lives, within the writings we have discussed theoretical positions are adopted which would lead to three differing demands over and above this general demand:

(1) that women should with the aid of science and technology abandon natural reproduction — essentially a theory of the technological fix (Firestone);

(2) that women should seek to control the management of birth-control technology, including its research and development (Dalla Costa); and

(3) that women should consort with men only to conceive (Rivolta Feminile).

SCIENCE FOR LIBERATION OR SUBJUGATION?

Firestone, having analysed the core of women's oppression as residing in her biological destiny, invokes the aid of science and technology to liberate her. For the social problems which confront woman (albeit with biological parameters), a technological fix is prescribed. Her conception of science is almost nineteenth-century in its confidence in the inherently progressive nature of science and technology. The naivete of this view is reinforced if we examine the social structure of science, for it is historically a stronghold of men and it is difficult to understand why science and technology should appear to be the allies of women. In a situation where not only the allocation of research funds between research areas substantially determines the structure of knowledge, but where the state is actively seeking greater control, where can Firestone's expectations of science seriously be placed?

It is not only that such hopes are unrealistic but also that their naivete about science may make it more difficult for women to see the dangers flowing from some new technologies of reproduction.

*The Ideologies of Regulating Reproduction*

The conflicting ideologies of birth control over the last century reflect the struggle over who shall control the technology, and at the same time reveal the nature of the technology itself. These issues, as we shall argue, are intensified by the proposed reproductive technologies of the future.

Regulating reproduction through late marriage, infanticide, abortion, some forms of circumcision, rudimentary chemical and physical intervention (vinegar on a sponge in the uterus, linen condoms) are, as Himes' classical text points out,[17] not new in medical history. However, over the last century science and technology have dramatically increased the efficacy of the means of regulating reproduction. Beginning with the advance in rubber technology in the 1870s, which produced more effective condoms and eventually the cap, research moved from physical methods of regulation to chemical methods. The triumph of the pill in the late 1950s marked a dramatic step foward for women, particularly in the industrialised societies. A return to physical means was evinced by the introduction of the coil, a technique which has been widely employed in developing societies.

Once the efficacy of scientific birth control was established, the ideological debate over its employment was also joined. The eugenicist movement, partly as an ideological expression of imperialism abroad and ruling-class interests at home, had long argued that population control of inferior people as defined by class and race was a necessary and desirable goal. Naturally it saw in scientific birth control a more acceptable way of securing its ends. The most brutal forms of eugenic control saw their apotheosis in Nazi Germany, beginning with sterilisation and marriage laws, but leading finally to the gas chamber. There were ripples in Britain in the proposals canvassed in the 1930s that the unemployed should be sterilised, and in the practice of shutting away in subnormal institutions young girls who had evinced their feeble-mindedness by having a baby without securing a husband first. Some such women were released only in 1973, having been shut away for thirty years for their 'crime'. In the United States, too, compulsory sterilisation was practised, particularly against poor blacks. Even in the 1970s there have been a continuing number of cases of the involuntary sterilisation of black women. In 1973 the American Civil Liberties Union took up the case of two young black sisters who, at twelve and fourteen, had been surgically sterilised in Montgomery, Alabama, in 1964.[18] It is also the situation that in some states free Medicare for women giving birth to a baby is dependent upon her agreeing to subsequent sterilisation. A similar situation seems in certain cities to be developing as hospital policy for abortions on the National Health Service.

Not surprisingly the eugenicist nature, both of internal programmes of birth control and also of those in developing countries (particularly when sponsored by imperialist countries) has been widely recognised by black and Third World movements. However, some of the debates which have taken place concerning population-control programmes show signs of acknowledging both the eugenicist element and also the liberatory element for women themselves.[19] Increasingly it seems that it is women who are recognising the positive aspect of birth-control technology even while sharing with men an agreement as to its societal function.

## Eco-eugenicists

To this traditional eugenicist argument has come a new ally, the ecologist movement preaching zero population growth. Starting from an unimpeachable position concerning the finite nature of the earth and its resources, it deftly moves into an argument for retaining the developing world as a natural wilderness, and restraining economic growth within the industrialised world. As little redistribution is proposed, zero population and the provision of wilderness seem to be little more than the claims of the 'haves' to preserve their advantages. Indeed the ecologist's interest in population control has mobilised a good deal of enthusiastic support from the middle classes, but ruling-class allies are drawn from groupings such as the Council for the Preservation of Rural England, itself representative of old landed and new business/landed interests — that is, the great bourgeoisie itself. In so far that women are made to feel guilty for polluting the earth with their children, the ecology movement merely adds sexism to the traditional racism and class dominance of the eugenicists' concerns.

## Social Engineering and Population Control

There is a long history of state attempts to control the population size. In feudal Japan, for example, in certain rural areas where infanticide has been carried out to such a level that the economic activity was threatened, the samurai provided family allowances to encourage the peasants to preserve their young. However, while the case for intervention was in this situation a matter of survival, intervention and estimates of a desired population level are rarely so straightforward; exhortation, financial inducements, enlarging or restricting birth-control facilities have all been used. France, for example, has had for many years a generous family-allowance programme and limited birth-control facilities but is still unable to increase her population to the levels deemed attractive by politicians.

Increasing the population through financial incentive is less efficient than the technique of first providing an extensive abortion service and then withdrawing it. The Soviet Union in the 1930s, Rumania in 1966 and Bulgaria, Czechoslavakia and Hungary in 1973 have all used this with some success. A decision which previously belonged to women is withdrawn — via the medical profession — to become the decision of the state.

Various social techniques are being used to reduce population. Negative family allowances, for example, where the mother receives family allowances only if she produces the right number of children, were discussed for Mauritius and have recently been adopted by Hong Kong. This, as studies of family budgeting under duress show, affects the fathers least because as the breadwinners they have to have food. It affects the children most sharply because of a young child's sensitivity to poor nutrition; together with the mothers because they typically deny themselves food in order to feed their children. Few administrative proposals for regulating population through economic incentives and disincentives have been so draconian, and it is not without significance that such a proposal was carried out in the context of a colony. Other techniques dependent on the provision of services to limit population are a mix of social and technological engineering — epitomised perhaps by the highly publicised and not very successful transistor/male sterilisation programme in India.

In Britain, although the birth rate is in fact slowing down — to some extent slackened by the increased provision of birth-control services and the liberalisation of the abortion law — there is a sustained attempt to generate an ideology of zero population growth. One recent conference of natural and social scientists went further, advocating a population for Britain of some forty million.[20]

Halperin, Kenrick and Segal[21] suggest that there are three possible ways of looking at the role and image of women in a situation in which women are no longer forced to produce children but are to be actively discouraged. The first is a continuation and intensification of the traditional view of women as the most flexible part of the industrial reserve army (last hired, first fired). The second relates to women not in the

paid labour force, or in the unpaid as producers of children; these need
to learn to accept themselves as 'super consumers' in the 'none is fun'
image. The third is to encourage women to take on wider social roles
and hence have fewer children, a view which echoes the Ross Report,[22]
which noted that zero population growth would be likely to result in
'increasing demands by married women for employment and other
opportunities for social participation outside the immediate family
circle'. In setting out these three roles the contradiction is revealed
between the first two and the third, which would entail a very different
kind of society. If a society, despite its zero population growth goal, is
unwilling to make significant social concessions to women, it must
resort to more repressive measures against them. But before we discuss
how this might be done, we will look at the liberatory potential in
developments in birth technology.

*Population Control in Women's Hands*

The area of folk medicine which related to childbirth, contraception
and abortion was traditionally, as Ehrenreich and English point out,[23]
in women's hands. The history of the professionalisation of medicine,
which occurred prior to the advent of scientific medicine, and was
therefore not primarily concerned with the replacement of superstition
and quackery, was also the history of the replacement of women
healers by male doctors. The attack on the 'witches' is interpreted by
Ehrenreich and English in a similar vein. They argue that the 'witches'
were practitioners of folk medicine who experienced civil repression at
the hands of the male-dominated Catholic Church precisely because
their work enabled women to control their reproductive functions. One
way, for example, of infallibly detecting a witch, cited by the *Malleus
Maleficorum*, was that the woman had been instrumental in securing an
abortion. Interference with the natural process of unlimited reproduction was thus the major reason for the Church's persecution of 'witches'.

The significance of women medical workers is seen too in the
twentieth century, when the potentiality of scientific birth control had
been realised.[24] The deep affection by women, particularly poor
women, for Marie Stopes and the doctors working in the pioneering birth-control clinics of the 1920s and 1930s stood in sharp contrast
to the feelings inspired by the eugenicist movement.[25] For that matter
the self-help centres springing up, first in the United States in the late
1960s and now in Britain, the pregnancy testing and abortion clinics
run by women for women, or the struggle to save the Elizabeth Garrett

Hospital for Women in London, reflect this growing conviction that the care of women's bodies should be in women's hands.

Yet both Marie Stopes and today's collectivist self-help centres are deviant strands within the general norms of a patriarchal and capitalist society, and we have to turn elsewhere, notably China, for a society which, having made some false starts on the planning of reproduction, has now established a line which both gives the control to women and is extremely successful. With the rapid population growth experienced since liberation, China has had to develop a clear policy on planning reproduction as much as on planning production. Thus, while some groups like the national minorities, after decades of population decline, are encouraged, through the introduction of fertility clinics, to have larger families, most, especially those in the economically abundant areas, are encouraged to limit their family size.

As Hawthorne points out, the number of children depends on seemingly small things such as the age people decide to get married. One key method is through encouraging relatively late marriage; in the towns the desired age is 25 for women and 27 for men, in the country 23 for women and 25 for men. After two children, spaced with the aid of the coil or pill, some women agree to be sterilised, a very final step for a woman who may be only 26. As Hans Suyin points out,[27] this programme of birth planning, the most efficient in the developing world, is carried out primarily by the women themselves, by word-of-mouth education and exemplary action. Even the barefoot doctors, like the more highly trained, play a relatively minor role, as 'women do not wish to be advised on these matters by young unmarried persons'. Scientific education and all poster display is eschewed as when it was tried the peasants felt that they were being invited to commit infanticide once more. Instead the women are encouraged through collective discussion to emulate the confident action of the women cadres, who, having had two children, have had themselves sterilised. The immense mobilisation of the women themselves discussing and deciding 'whose turn it is to have a baby' means that the interests of the whole society and those of the individual women are seen as harmonious.

BIOLOGICAL ENGINEERING AND THE MANAGED SOCIETY

However for Western capitalist society, the commitment to individualism excludes these collectivist solutions, and the relatively modest success of the existing array of social engineering strategies leads, if the state requires more population control, to one which is primarily biological engineering. It is therefore to some of the new technologies at least partially under development to which we now turn our attention.

*Choosing Boys with Etzioni*

One of the possibilities which is well advanced technically concerns the choice of the sex of the unborn child. Despite the cultural-lag hypothesis of the time gap between technological change and societal adaptation, only Amitai Etzioni[28] seems as a sociologist to have given any thought to the kind of society this intervention in reproduction will produce. Sharing the assumption that Western civilisation proceeds by adjusting to new technologies, Etzioni considers the implication of choosing boy babies. People he notes, have measurable preferences about the sex of their children. More families stop having children after a boy is born than a girl, and men, in particular, state a preference for boys. He predicts that if it were actually possible to choose, many more than those who stated a preference when there was no means of definitely acquiring it would choose boys, 'even if it were taboo or unpopular at first . . . [it] could become quite widely practised once it became fashionable', thus defining  the role for advertising companies. In a capitalist society nothing is more sacrosanct than the 'individual choice' of the consumer, thus making any move to prevent such techniques from being commercially developed much more difficult. There is an echo of this in the debate over the production of test-tube babies announced at the 1974 British Medical Association meeting by Professor Bevis. While traditional morality questions whether such production is ethically permissible, the doctors legitimise themselves by invoking the individual choice of the mothers themselves. However, at the level of individual choice the ability to determine the sex of an unborn child has more market potential than choosing to have a test-tube baby. There would be no difficulty in securing investment to launch a sex determinant as it would be obviously very profitable.

Apart from the logistics of securing the production and adoption of the sex determinant, Etzioni permits himself to consider the societal effect. Clearly a sexual imbalance would occur, though not an enormous imbalance, and therefore homosexuality would increase. The shortage of women would also increase female prostitution and there would be certain changes in cultural life. Over all, in terms of the basic power and economic structures of society, Etzioni thinks the new sexual imbalance would have little effect. The society he portrays is in fact partially observable in those towns and areas in Western Europe where single male migrant workers are employed, although achieved in this case through economic pressures. This type of development strengthens the existing social order, while worsening conditions for those in the lower strata.

### 'Leaping to Breed Male' with Postgate

The biologist Postgate[29] is a bolder sociological prophet. Firestone's antithesis, he offers a male Utopian conception of how sex selection would change male – female relations and deal with the population question. Like Firestone's, Postgate's thesis depends on a technological fix. Unlike Firestone he, as a biologist, is part of that fix.

Postgate links the three eugenicist concerns; class, race and sex. He argues that overpopulation is the most important problem facing humanity today, and that starvation and social instability are dependent upon it. Birth control is rejected as he claims it works best in countries that least need it, namely 'wealthy educated countries' but not in 'underdeveloped unenlightened ones'. Alternative forms of population control, such as war, disease, legalised infanticide, euthanasia, are rejected as they are not selective, acceptable, quickly effective or permanent enough.

Breeding male, however, meets these criteria, as 'countless millions of people would leap at the opportunity to breed male' (particularly in the third world), and 'no compulsion or even propaganda would be needed to encourage its use, only evidence of success by example'. The rate of population decline brought about by the fact of fewer females would depend on how quickly 'ordinary' men and women catch on to what is happening. When the world population stabilised at a much lower number of people, the real benefits of developing changes in industrial production could be reaped.

Some people, of course, might object to a 'male-child pill' because of the social consequences of the 'transitional phase', which he described as a 'matter of taste, rather than serious concern'. During this period women would be kept in purdah, no longer able to work or travel freely, given as rewards for the most outstanding males; polyandry could be introduced, and women come to be treated as queen ants. While no woman could begin to equal a queen ant in production, the term indicates an intensified technological process where women bear children throughout their fertile years; the analogy is factory farming; battery rather than free-range hens. Thus the production of children would mirror existing industrial techniques; a realistic method once the absolute numbers of women are reduced. The logic of capitalism is to rationalise production, and this is one way it could be accomplished. Rowbotham's suggestion that the logical thrust of 'pure capitalism' is towards 'baby farms and state controlled breeding' is another.[30]

While Postgate's utopia will come to resemble (we quote) 'a giant boys public school or huge male prison', the model the physicist Shockley offers us is one of friendly fascism. (Postgate's view can perhaps be compared with National Socialism before the Night of the Long Knives, while Shockley's is after.

## Friendly Fascism with Shockley

Shockley, Nobel Laureate for his work on the transistor, is also concerned with the population question. Although his plan embraces both social and biological engineering, it turns on a new technology and is therefore included here.[31] He offers a population-control plan that reinforces the monogamous family unit, and strengthens the patriarchal system. There are five steps, beginning by convincing people that population limitation is desirable and necessary for survival. The Census Bureau then calculates the number of children each woman may have (2·2 if an 0·3 per cent increase is permitted each year). The Public Health Department then sterilises every girl as she enters puberty by a subcutaneous injection of a contraceptive time capsule which provides a slow seepage of contraceptive hormones until it is removed. When the girl marries she is issued twenty-two deci-child certificates, and on payment of ten certificates her doctor will remove the contraceptive capsule, which is replaced when the baby is born. After two babies, the couple may either sell their remaining two certificates (through the Stock Exchange), or try to buy eight more on the open market and have a third child. Those who do not have children have twenty-two to sell.

This technology, which in principle is not impossible, transfers control not merely from the woman to the man — as in the case of such techniques as the condom — but directly to the state. Nor is it chance that Shockley's baby credit system would disfavour the low income groups, which in the United States would felicitously coincide with the Blacks, a group he regards as eugenically inferior. Low-income people are more likely to sell their credits, and in this friendly model the eugenicist goal of limiting inferior children is achieved by the workings of the market. In addition, as power is not distributed evenly within the family unit, the situation of all women, but particularly those with low incomes, who may want children when their husbands do not, would be markedly worsened.

## Cloning Geniuses — Fact or Fantasy

Cloning is asexual reproduction of identical individuals. One technique in animals is to transplant the nucleus of any body cell into the denucleated infertile female egg cell. This has formed a central element in the proof that each body cell contains all the genetic information originally present in the fused sperm and ovum. Asexual reproduction is different from sexual reproduction not only in its process but also in that the outcomes are different. Cloning produces, or rather potentially reproduces, large numbers of identical people. The work, originally carried out on frogs, is thus central to the theories of developmental biology. What we have recently seen is the transfer of this work to human genetics, ostensibly because knowledge of foetal growth would aid the clinical treatment of infertility. The best known of this work on test-tube babies has been carried out in Britain by Edwards, Bannister and Steptoe[32] and is conducted within a particular set of beliefs about women and their biologically determined social role. Like Firestone, Edwards and Steptoe recognise the significance of biological reproduction in the structure of the family. Their work is explicitly designed to enable women, otherwise barren, to fulfill their biological destiny. They set to one side the possibility of adoption. Thus their work, using the language of helping women, is in fact deeply conservative in terms of preserving the woman's role. Their speculation that some women, who are better breeders, could carry the babies of other mothers, suggests a form of biological emancipation of a dominant class of women achieved only by the biological exploitation of another subordinate class. Updated and appropriately modernised by the scientific and technological revolution, it is at root uncommonly similar to wet nursing, whereby poor women nursed rich women's babies at the expense of their own children and bodies. 'Rent-a-tit' gives way to 'rent-a-belly', and it is suggested that this is progress.

Apart from the particular implications for women, the general and sinister implications of cloning have not escaped a public educated by Aldous Huxley's Brave New World, where natural reproduction had become obsolete and babies were bred in test tubes to be conditioned for their place in the division of labour by the best behaviourist techniques.

What is interesting in this situation is that where in the 1930s the Marxist geneticists, such as J.B.S. Haldane[33] or H. Muller[34] considered human cloning to be an interesting and eventually practical possibility, today the orthodox left liberal geneticists such as J. Lederberg, M. Pollock, W. Hayes and M.H. Wilkins seem to suggest that cloning is a science fiction, which, because it is unthinkable, will therefore not happen.[35] It remains for the right to argue that cloning is practicable. J.D. Watson (like Wilkins one of the DNA model-builders and thus a leading spokesman for the new biology argued in *Atlantic Monthly* that 'We must therefore assume that techniques for the *in vitro* manipulation of human eggs are likely to be in general practice, capable of routine performance in many major countries, within some ten to twenty years'.

The possibilities of cloning for genius nourish the strong eugenicist current which has emerged over the last five years or so. Mainly focused around the IQ question, the scientific legitimation for contemporary eugenicism is provided by the work of W. Shockley, A.R. Jensen, H.J. Eysenck and R. Herrnstein. For the most part the response to this latest wave of eugenicism has raised questions concerning class and race; to this we would add the third dimension — sex.

If the discussion of cloning is carried out at a biological level, then because the women has both the egg cells — to be denucleated — and the body cells from which the nucleus can be transferred to the egg cells, in principle a kind of virgin reproduction could take place. In this sense a world without men is biologically possible, and, further, so long as female egg cells are necessary, a world without women is not (although as each woman has several hundred egg cells, it would be possible to make do with very few women).

The main argument against the advocacy of cloning for the women's movement must be that it mystifies the actual character of science and technology under capitalism. When we examine the kinds of people who might contribute the crucial nuclei of their body cells discussed by the potential engineers themselves, we find that they are males and are characterised by cerebral and power attributes (for example Lenin — as in Müller's original scheme — or Einstein — a favourite candidate of both left and right scientists) whereas women are chosen for their sexual attraction. The only two names of women we have been able to find considered by the genetic engineers as suitable for cloning are Brigitte Bardot and Elizabeth Taylor.

Thus Firestone, as one of the most influential of the radical feminists, has, in her misplaced advocacy of asexual modes of reproduction, ignored the actual character of science and technology under contemporary capitalism. Essentially her argument is that of the *deus ex machina*, whereby the (male) god in a white laboratory coat resolves the biologically founded contradictions of the woman's situation. The male biological engineers are thus to create the feminist Utopia.

The Italian radical feminists, while acknowledging that reproduction is central, avoid the trap of Firestone's technological fix, instead arguing that women should only consort to conceive. It is left to Dalla Costa and James to argue that women must work to control the means of regulating reproduction, including the research and development of new technologies. However as the proposals are seen by the biological engineers themselves, there is little advantage to women, whether the proposal is choosing the sex of the unborn child — almost immediately practicable under liberal capitalism — or whether it involves 'leaping to breed male', or quasi-sterilisation, or cloning, which would require a more corporate form of capitalism for successful implementation.

Between the old technologies of regulating reproduction and the new is a major shift from the individually controlled technology to a potentially state-controlled technology. Between using a condom or a pill and putting a contraceptive agent in the drinking water lies not merely a comparison of lesser or greater efficiency in birth control, but a whole series of complex issues concerning social relationships, the relationship of women to their own bodies, and so forth.

Women's inferior social position both in terms of status and in terms of power will make it easier to deprive them of even their present gains in terms of controlling their reproductive function. Further, if the past holds a key for understanding the future, women are in for a very tough time indeed.[36] Rather than forcing reproduction upon her as did the Catholic Church, today's religion, science, seems to be moving in the opposite direction, limiting her right to reproduce, if not asking her to commit 'voluntary' femicide.

Technology is thus not something separate from humanity, but as Marx puts it, 'technology discloses man's mode of dealing with nature, the process of production by which he sustains his life, and thereby also lays bare the mode of formation of his social relations and of the mental conceptions that flow from them'.[37] If this view of technology is applied to birth control, potentially reproduction could be transformed by being for all women a voluntary act. Women could live in harmony with their biology. Nature would be humanised and the long struggle from nature to a truly human culture advanced. But while it becomes possible through science and technology to humanise nature, dehumanising nature by turning reproduction into a mechanical process is a logical development within an increasingly managed society.

# 9

# A Critique of Political Ecology

*Hans Magnus Enzensburger*

As a scientific discipline, ecology is almost exactly a hundred years old. The concept emerged for the first time in 1868 when the German biologist, Ernst Haeckel, in his *Natural History of Creation*, proposed giving this name to a sub-discipline of zoology — one which would investigate the totality of relationships between an animal species and its inorganic and organic environment. Compared with the present state of ecology, such a proposal suggests a comparatively modest programme. Yet none of the restrictions contained in it proved to be tenable: neither the preference given to animal species over plant species, nor to macro- as opposed to micro-organisms. With the discovery of whole ecosystems, the perspective which Haeckel had had in mind became redundant. Instead there emerged the concept of mutual dependence and of a balance between all the inhabitants of an ecosystem, and in the course of this development the range and complexity of the new discipline have grown rapidly. Ecology became as controversial as it is today only when it decided to include a very particular species of animal in its researches — the human. While this step brought ecology unheard of publicity it also precipitated it into a crisis about its validity and methodology, the end of which is not yet in sight.

Human ecology is, first of all, a hybrid discipline. In it categories and methods drawn from the natural and social sciences have to be used together, without this in any way theoretically resolving the resulting complications. Human ecology tends to suck in more and more new disciplines and to subsume them under its own research aims. This tendency is justified not on scientific grounds but because of the urgency of ecology's aims. Under the pressure of public debate ecology's statements in recent years became more and more markedly prognostic. This 'futurological deformation' was totally alien to ecology

so long as it considered itself to be merely a particular area of biology.
It must be clearly understood that this science has now come to lay
claim to a total validity — a claim which it cannot make good. The
more far-reaching its conclusions, the less reliable it is. Since no one
can vouch for the accuracy of the enormous volume of material from
every conceivable science on which its hypotheses are constructed, it
must — precisely to the degree that it wishes to make global statements
— confine itself to working syntheses. One of the best known ecological
handbooks -- *Population, Resources, Environment* by Paul and Anne
Ehrlich — deploys evidence from the following branches of sciences,
either implicitly or explicitly: statistics, systems theory, cybernetics,
games theory and prediction theory; thermodynamics, biochemistry,
biology, oceanography, mineralogy, meteorology, genetics; physiology,
medicine, epidemology, toxicology; agricultural science, urban studies,
demography; technologies of all kinds; theories of society, sociology
and economics (the latter admittedly in a most elementary form). The
list is not complete. It is hard to describe the methodological confusion
that results from the attempt at a synthesis of this sort. If one starts
from this theoretical position there can, obviously, be no question of
producing a group of people who are competent to deal with it. From
now on ecology is marginally relevant to everyone; and this, incidentally, is what makes the statements in this chapter possible.

THE CENTRAL HYPOTHESIS

What till recently was a marginal science has within a few years become
the centre of bitter controversies. This cannot be explained merely by
the snowballing effect of the mass media. It is connected with the central statement made by human ecology — a statement that refers to the
future and is therefore at one and the same time prognostic and hypo-
thetical. On the one hand, everyone is affected by the statement, since
it relates to the existence of the species; on the other, no one can form
a clear and final judgement on it because, in the last resort, it can only
be verified or proved wrong in the future. This hypothesis can be formulated as follows: the industrial societies of this earth are producing
ecological contradictions, which must in the foreseeable future lead to
their collapse.

In contradistinction to other earlier theories of catastrophe this
prognosis does not rest on linear, monocausal arguments. On the contrary, it introduces several synergistic factors. A very simplified list
of the different chains of causality would look something like this:

(1) Industrialisation leads to an uncontrolled growth in world population. Simultaneously the material needs of that population increase. Even given an enormous expansion in industrial production, the chances of satisfying human needs deteriorate *per capita*.

(2) The industrial process has up to now been nourished from sources of energy which are not in the main self-renewing: among these are fossil fuels as well as supplies of fissile material like uranium. In a determinable space of time these supplies will be exhausted; their replacement through what are basically new sources of energy (such as atomic fusion) is theoretically conceivable, but not yet practically realisable.

(3) The industrial process is also dependent on the employment of mineral raw materials — above all of metals — which are not self-renewing either; their exploitation is advancing so rapidly that the exhaustion of deposits can be foreseen.

(4) The water requirements of the industrial process have reached a point where they can no longer be satisfied by the natural circulation of water. As a result, the reserves of water in the ground are being attacked; this must lead to disturbances in the present cycle of evaporation and precipitation and to climatic changes. The only possible solution is the desalination of sea-water; but this is so energy-intensive that it would accelerate the process described in (2) above.

(5) A further limiting factor is the production of foodstuffs. Neither the area of land suitable for cultivation nor the yield per acre can be arbitrarily increased. Attempts to increase the productivity of farming lead, beyond a certain point, to new ecological imbalances, for example erosion, pollution through poisonous substances, reductions in genetic variability. The production of food from the sea comes up against ecological limits of another kind.

(6) A further factor — but only one factor among a number of others — is the notorious 'pollution' of the earth. This category is misleading in so far as it presupposes a 'clean' world. This has naturally never existed and is moreover ecologically neither conceivable nor desirable. What is actually meant are disequilibriums and dysfunctionings of all kinds in the metabolism between nature and human society occurring as the unintentional side-effects of the industrial process. The polycausal linking of these effects is of unimaginable complexity. Poisoning caused by harmful substances — physiological damage from pesticides, radio-active isotopes, detergents, pharmaceutical preparations, food additives, artificial manures, trace quantities of lead and mercury, fluoride, carcinogens, mutant genes, and a vast quantity of other substances, are

only one facet of the problem. The problem of irreversible waste is only another facet of the same question. The changes in the atmosphere and in the resources of land and water traceable to metabolic causes such as production of smog, changes in climate, irreversible changes to rivers and lakes, oceanographic changes, must also be taken into account.

(7) Scientific research into yet another factor does not appear to have got beyond the preliminary stages. There are no established critical quantifications of what is called 'psychic pollution'. Under this heading come: increasing exposure to excessive noise and other irritants, the psychical effects of overpopulation, as well as other stress factors which are difficult to isolate.

(8) A final critical limit is presented by 'thermal pollution'. The laws of thermodynamics show that, even in principle, this limit cannot be crossed. Heat is emitted by all processes involving the conversion of energy. The consequences for the global supply of heat have not been made sufficiently clear.

A basic difficulty in the construction – or refutation – of ecological hypotheses is that the processes invoked do not take place serially but in close interdependence. That is also true of all attempts to find solutions to ecological crises. It often, if not always, emerges that measures to control one critical factor lead to another getting out of control. One is dealing with a series of closed circuits, or rather of interference circuits, which are in many ways linked. Any discussion that attempted to deal with the alleged 'causes' piecemeal and to disprove them singly would miss the core of the ecological debate and would fall below the level which the debate has meantime reached.[1]

Yet even if there exists a certain, but by no means complete, consensus that the present process of industrialisation must lead *ceteris paribus* to a breakdown, three important questions connected with the prognosis are still open to debate. The first concerns the time scale involved. Estimations of the point in time at which a galloping deterioration of the ecological situation may be expected differ by a magnitude of several centuries. They range from the end of the 1980s to the twenty-second century. In view of the innumerable variables involved in the calculations, such divergencies are not to be wondered at. (For example the critics of the MIT report, *The Limits of Growth*, have objected to the results given there on the grounds that the mathematical model on which it is based is much too simple and the number of variables too limited.) A second controversial point is closely related to the first; namely that the relative weighting to be given to the individual factors which are blamed for the catastrophe is not made clear. This is a point at issue, for example, in the debate between Barry Commoner and Paul Ehrlich. While the latter considers population growth to be the 'critical factor', the former believes that the decisive factor is modern industrial technology. An exact analysis of the factors involved comes up against immense methodological difficulties. The scientific debate between the two schools therefore remains undecided.

Third, it is obviously not clear what qualifies as an environmental catastrophe. In this connection one can distinguish a number of different perspectives dictated by expectation or fear. There are ecologists who concern themselves only with mounting dangers and the corresponding physiological, climatic, social and political 'disturbances'; others, like the Swedish ecologist, Gösta Ehrensvärd, contemplate the end of social structures based on industrialisation; some prognoses go further — those of what in the United States are called 'doomsters' talk of the dying out of the human species or the disappearance from the planet of a whole series of species — primates, mammals and vertebrates. The tone in which the respective ecological hypotheses are presented ranges correspondingly from the mildest reformist warnings to deepest resignation. What is decisive for the differences between them is naturally the question of how far the process of ecological destruction and uncontrolled exploitation is to be regarded as irreversible. In the literature, the answer to this question is made to depend on the one hand on an analysis of the factors involved; on the other, on temporal parameters. The uncertainty which is admitted to prevail on these two points means that there is no prospect of a firm answer. Authors like Ehrensvärd, who start from the premise that the end of industrial societies is at hand, and are already busy with preparations for a post-industrial society — one which, it should be added, contains a number of idyllic traits — are still in the minority. Most ecologists imply that they consider that the damage done so far is reversible, if only by tacking on to their analyses proposals to avert the catastrophe of which they are the prophets. These proposals will need to be critically examined.

THE ECOLOGICAL 'MOVEMENT'

Ecology's hypotheses about the future of industrialisation have been disseminated, at least in industrialised capitalist countries, through the mass media. The debate on the subject has itself to some extent acquired a mass character, particularly in the Anglo-Saxon and Scandinavian countries. It has led to the rise of a wide, although loosely organised, movement whose political potential is hard to estimate. At the same time the problem under discussion is peculiarly ill-defined. Even the statements of the ecologists themselves alternate between the construction of theories and broad statements of *Weltanschauung*, between precise research and totalising theories linked to the philosophy of history. The thinking of the ecological groups therefore gives the impression of being at once obscure and confused. The very fact that it is disseminated by the mass media means that the debate generally loses a great deal of its stringency and content. Subordinate questions such as that of recycling refuse or 'pollution' are treated in isolation; hypotheses are presented as certain spectacular cases of poisoning are sensationally exploited: isolated results of research are given absolute validity, and so on. Processing through the sewage system of industrialised publicity has therefore, to some extent, led to further pollution of a cluster of problems which from the start cannot be presented in a 'pure' way. This lack of clarity is propagated in the groups which are at present actively occupied with the subject of ecology, or rather with its *disjecta membra*, with what is left of it. The most powerful of these groups is that of the technocrats, who, at all levels of the state apparatus and also of industry, are busy finding the speediest solutions to particular problems – 'quick technological fixes' – and implementing them. This they do whenever there is a considerable potential for economic or political conflict – and only then. These people consider themselves to be entirely pragmatic – that is to say, they are servants of the ruling class at present in power – and cannot be assumed to have a proper awareness of the problem. They can be included in the ecological movement only in so far as they belong – as will be demonstrated – to its manipulators and in so far as they benefit from it. The political motives and interests in these cases are either obvious – as with the Club of Rome, a consortium of top managers and bureaucrats – or can easily and unequivocally be established.

What is less unequivocal is the political character of a second form of ecological awareness and the practice that corresponds to it. Here it is a matter of smaller groups of 'concerned and responsible citizens', as they say in the United States. The expression points, as does its German parallel, 'citizen's initiative', to the class background of those involved in it. They are overwhelmingly members of the middle class, and of the new petty bourgeoisie. Their activities have generally modest goals.

They are concerned with preserving open spaces or trees. Classes of schoolchildren are encouraged to clean up litter on beaches or recreation grounds. A boycott of non-decaying packaging is organised, and so on. The harmless impression made by projects of this kind can easily blind us to the reserves of militancy which they conceal. There only needs to be a tiny alteration in the definition of goals and these groups spontaneously begin to increase in size and power. They are then able to prevent the carrying through of large-scale projects like the siting of an airport or an oil refinery, to force high-tension cables to be laid underground or a motorway to be diverted. But even achievements of this magnitude only represent the limits of their effectiveness for a time. If the hypotheses of the ecologists should come even partially true, the ecological action groups will become a force of the first order in domestic politics and one that can no longer be ignored. On the one hand, they express powerful and legitimate needs of those who engage in these activities; on the other hand, they set their sights on immediate targets, which are not understood politically, and incline to a kind of indulgence in social illusion. This makes them ideal fodder for demagogues and interested third parties. But the limited nature of their initiatives should not conceal the fact that there lies within them the seed of a possible mass movement.

Finally, there is that part of the ecological movement which considers itself to be its hard core but which, in fact, plays a rather marginal role. These are the 'eco-freaks'. These groups, which have mostly split off from the American protest movement, are engaged in a kind of systematic flight from the cities and from civilisation. They live in rural communes, grow their own food, and seek a 'natural way of life', which may be regarded as the simulation of pre- or post-industrial conditions. They look for salvation in detailed, precisely stipulated dietary habits — eating 'earth food' — and agricultural methods. Their class background corresponds to that of the hippies of the 1960s — of reduced middle-class origin, enriched by elements from peripheral groups. Ideologically they incline towards obscurantism and sectarianism.

On the whole one can say that in the ecological movement — or perhaps one should say movements — the scientific aspects, which derive predominantly from biology, have merged in an extremely confused alliance with a whole series of political motivations and interests, which are partly manifest, partly concealed. At a deeper level one can identify a great number of socio-psychological needs, which are usually aroused without those concerned being able to see through them. These include: hopes of conversion and redemption, delight in the collapse of things, feelings of guilt and resignation, escapism and hostility to civilisation.

In these circumstances it is not surprising that the European left holds aloof from the ecological movement. It is true that it has incorporated certain topics from the environmental debate in the repertory of its anti-capitalist agitation; but it maintains a sceptical attitude to the basic hypothesis underlying ecology and avoids entering into alliances with groups which are entirely orientated towards ecological questions. The left has instead seen its task to be to face the problem in terms of an ideological critique. It therefore functions chiefly as an instrument of clarification, as a tribunal which attempts to dispel the innumerable mystifications which dominate ecological thinking and have encouraged it. The most important elements in this process of clarification, which is absolutely necessary, are listed and discussed below.

## THE CLASS CHARACTER OF THE CURRENT ECOLOGICAL DEBATE

The social neutrality to which the ecological debate lays claim, having recourse as it does so to strategies derived from the evidence of the natural sciences, is a fiction. A simple piece of historical reflection shows just how far this class neutrality goes. Industrialisation made whole towns and areas of the countryside uninhabitable as long as a hundred and fifty years ago. The environmental conditions at places of work, that is to say in the English factories and pits, were — as innumerable documents demonstrate — dangerous to life. There was an infernal noise; the air people breathed was polluted with explosive and poisonous gases as well as with carcinogenous matter and particles which were highly contaminated with bacteria. The smell was unimaginable. In the labour process contagious poisons of all kinds were used. The diet was bad. Food was adulterated. Safety measures were non-existent or were ignored. The overcrowding in the working-class quarters was notorious. The situation over drinking water and drainage was terrifying. There was in general no organised method for disposing of refuse:

> when cholera prevailed in that district [Tranent, in Scotland] some
> of the patients suffered very much indeed from want of water, and
> so great was the privation, that on that calamitous occasion people
> went into the ploughed fields and gathered rain water which collec-
> ted in depressions in the ground, and actually in the prints made by

horses' feet. Tranent was formerly well-supplied with water of excellent quality by a spring above the village, which flows through a sand-bed. The water flows into Tranent at its head . . . and is received into about ten wells, distributed throughout the village. The people supply themselves at these wells when they contain water. When the supply is small, the water pours in a very small stream only. . . . I have seen women fighting for water. The wells are sometimes frequented throughout the whole night. It was generally believed by the population that this stoppage of the water was owing to its stream being diverted into a coal-pit which was sunk in the sand-bed above Tranent.[2]

These conditions, which are substantiated by innumerable other sources from the nineteenth century, would undoubtedly have presented a 'neutral observer' with food for ecological reflection. But there were no such observers. It occurred to no one to draw pessimistic conclusions about the future of industrialisation from these facts. The ecological movement has only come into being since the districts which the bourgeoisie inhabit and their living conditions have been exposed to those environmental burdens that industrialisation brings with it. What fills their prophets with terror is not so much ecological decline, which has been present since time immemorial, as its universalisation. To isolate oneself from this process becomes increasingly difficult. It deploys a dialectic which in the last resort turns against its own beneficiaries. Pleasure trips and expensive packaging, for example, are by no means phenomena which have emerged only in the last decades; they are part of the traditional consumption of the ruling classes. They have become problematic, however, in the shape of tourism and the litter of consumerism; that is, only since the labouring masses have shared them. Quantitative increase tips over into a new quality — that of destruction. What was previously privilege now appears as nightmare and capitalist industry proceeds to take tardy, if still comparatively mild, revenge on those who up to now had only derived benefit from it. The real capitalist class, which is decreasing in numbers, can admittedly still avoid these consequences. It can buy its own private beaches and employ lackeys of all kinds. But for both the old and the new petty bourgeoisie such expenditure is unthinkable. The cost of a private 'environment' which makes it possible to escape to some extent from the consequences of industrialisation is already astronomical and will rise more sharply in future.

It is after all easy to understand that the working class cares little about general environmental problems and is only prepared to take part in campaigns where it is a question of directly improving their working and living conditions. In so far as it can be considered a source of ideology, ecology is a matter that concerns the middle class. If avowed representatives of monopoly capitalism have recently become its spokesmen — as in the Club of Rome — that is because of reasons which have little to do with the living conditions of the ruling class. These reasons require analysis.

## THE INTERESTS OF THE ECO-INDUSTRIAL COMPLEX

That the capitalist mode of production has catastrophic consequences is a commonplace of Marxism, which also not infrequently crops up in the arguments of the ecological movement. Certainly the fight for a 'clean' environment always contains anti-capitalist elements. Nevertheless fascism in Germany and Italy has demonstrated how easily such elements can be turned round and become tools in the service of the interests of capital.[3] It is therefore not surprising that ecological protest, at least in Western Europe, almost always ends up with an appeal to the state. Under present political conditions this means that it appeals to reformism and to technocratic rationality. This appeal is then answered by government programmes which promise an 'improvement in the quality of life', without of course indicating whose life is going to be made more beautiful, in what way and at whose expense. The state only

> goes into action when the earning powers of the entrepreneur are threatened. Today the environmental crisis presents a massive threat to these interests. On the one hand it threatens the material basis of production — air, earth and water — while on the other hand it threatens man, the productive factor, whose usefulness is being reduced by frequent physical and psychical illnesses.[4]

To these have to be added the danger of uncontrollable riots over ecological questions as the conditions in the environment progressively deteriorate.

On the question of state intervention and 'environmental protection from above', the Left's ideological critique displays a remarkable lack of historical reflection. Here too it is certainly not a question of new phenomena. The negative effects of environmental damage on the earning power of industry, the struggle over the off-loading of liability, over laws relating to the environment and over the range of state control can be traced back without much difficulty to the early period of British industrialisation; a remarkable lack of variation in the attitude of the interests involved emerges from such a study. The previously quoted report on the water supply and the drainage problems in a Scottish mining village is taken from an official report of the year 1842 — one which incidentally was also quoted by Engels in his book on *The Condition of the English Working Class*. The chairman of the commission of inquiry was a certain Sir Edwin Chadwick, a typical predecessor of the modern ecological technocrats. Chadwick was a follower of the utilitarian political philosopher and lawyer, Jeremy Bentham, of whom Marx said: 'If I had the courage of my friend H. Heine, I would call Mr. Jeremiah a genius at bourgeois stupidity.'[5] James Ridgeway, one of the few American ecologists capable of intervening in the present environmental discussion with political arguments, has dealt thoroughly with Chadwick's role.[6] Then as now the rhetoric of the ecological reformers served to cloak quite concrete connections between a variety of interests. The technological means with which this 'reform from above' operates have also altered less than one might think.[7]

But an historical perspective fails in its object if it is used to reduce modern problems to the level of past ones. Ridgeway does not always avoid this danger: he tends to restrict himself to traditional ecological questions like water pollution and the supply of energy. Without meaning to do so he thereby reduces the extent of the threatened catastrophe. It is true that there were environmental crises before this and that the mechanisms of reformist managements set up to deal with the crises can look back on a long history. What has to be kept in mind, however, is that the ecological risks have not only increased quantitatively but have taken on a new quality.

In line with the changes which have taken place in the economic base this also holds true for environmental pollution and state intervention. In its present form monopoly capitalism is inclined, as is well known, to solve its demand problems by extravagant expenditure at the cost of the public exchequer. The most obvious examples of this are unproductive investment in armaments and in space exploration. Industrial protection of the environment emerges as a new growth area the costs of which can either be off-loaded on to prices, or are directly made a social charge through the budget in the form of subsidies, tax concessions, and direct measures by the public authorities, while the profits accrue to the monopolies: 'According to the calculations of the American Council of Environmental Quality at least a million dollars is pocketed in the course of the elimination of three million dollars worth of damage to the environment.'[8]

Thus the recognition of the problems attendant on industrial growth serves to promote a new growth industry. The rapidly expanding eco-industrial complex makes profits in two ways: on the straightforward market, where consumer goods for private consumption are produced with increasing pollution, and in another where that same pollution has to be contained by control techniques financed by the public. This process at the same time increases the concentration of capital in the hands of a few international concerns, since the smaller industrial plants are not in the position to provide their own finance for the development of systems designed to protect the environment.

For these reasons the monopolies attempt to acquire influence over the ecological movement. The MIT study commissioned by the Club of Rome is by no means the only initiative of this kind. The monopolies are also represented in all state and private commissions on the protection of the environment. Their influence on legislation is decisive, and there are numerous indications that even apparently spontaneous ecological campaigns have been promoted by large firms and government departments. There emerges a policy of 'alliances from above', whose demagogic motives are obvious.[9]

By no means all ecological movements based on private initiative put themselves at the service of the interests of capital with such servility. That is demonstrated by the fact that their emergence has often led to confrontations with the police. The danger of being used is, however, always present. It must also be remembered that the interests of capital contain their own contradictions. Ecological controversies often mirror the clash of interests of different groups of entrepreneurs without their initiators always being clear as to the stakes involved in the campaigns. A long process of clarification will be necessary before the ecological movement has reached that minimum degree of political consciousness which it would require finally to understand who its enemy is and whose interests it has to defend.[10]

## DEMOGRAPHY AND IMPERIALISM

Warnings about the consequences of uncontrolled population growth — the so-called population explosion — also contain ideological motives and behind the demands to contain it are concealed political interests which do not reveal themselves openly. The neo-Malthusian arguments which authors like Ehrlich and Taylor have been at pains to popularise found expression at a particular moment in time and in a quite particular political context. They originate almost exclusively from North American sources and can be dated to the late 1950s and early 1960s — a time, that is to say, when the liberation movements in the Third World began to become a central problem for the leading imperialist power. (On the other hand the rate of increase in population had begun to rise much earlier, in the 1930s and 1940s.)

That this is no mere coincidence was first recognised and expressed by the Cubans:

At that time (1962) the Population Council in New York, supported by the Population Reference Bureau Inc. in Washington, launched an extensive publicity campaign for neo-Malthusianism with massive financial help from the Ford and Rockefeller Foundations, which contributed millions of dollars. The campaign pursued a double goal, which may even be attained: the ruling classes of Latin America were to be persuaded by means of skilful propaganda based on the findings of the FAO and work done by numerous, even progressive scientists, that a demographic increase of 2·5 per cent in Latin America would lead to a catastrophe of incalculable dimensions. The following excerpts from the report of the Rockefeller Foundation for 1965 are typical of this literature *made in the USA*: 'The pessimistic prediction that humanity is soon likely to be stifled by its own growth increasingly confronts all attempts to bring about an improvement in living standards. . . . It is clear that mankind will double in numbers in the life-time of two generations unless the present growth tendency is brought under control. The results will be catastrophic for innumerable millions of individuals'. The Population Reference Bureau expresses itself even more unequivocally: 'The future of the world will be decided in the Latin American continent, in Asia and Africa, because in these developing territories the highest demographic rates of growth have been registered. Either the birth rates must be lowered or the death rate must rise again if the growth is to be brought under control. . . . . The biologists, sociologists and economists of the Bureau have forecast the moment when Malthus' theory will return like a ghost and haunt the nations of the earth. (P.R.B. press statement of October 1966.)

The Cuban report also quotes Lyndon B. Johnson's remark to the effect that 'five dollars put into birth control is more useful in Latin America than a hundred dollars invested in economic growth.'[11] It adds: 'A comment on this cynical statement seems to us to be superfluous.'

Indeed not much intelligence is needed to discover behind the benevolent pose of the Americans both strong political motivation and the irrational fears which are responsible for the massive attempt by official and private groups in the United States to export birth control to the countries of the Third World. The imperialist nations see the time coming when they will be only a small minority when compared to the rest of the world and their governments fear that population pressures will become a source of political and, in the last analysis, military power. Admittedly fears of another kind can be detected underneath the rational calculations: symptoms of a certain panic, the precursors of which are easily recognisable in history. One has only to think of the hysterical slogans of the heyday of imperialism — 'the Yellow Peril' — and of the period of German fascism — 'the Red Hordes'. The 'politics' of population have never been free of irrational and racist traits; they always contain demagogic elements and are always prone to arouse atavistic feelings. This is admittedly true not only for the imperialist side. Even the Cuban source does not stop at the extremely enlightening comment that has been quoted but continues as follows:

> Fidel Castro has spoken on the question many times. We recall his words now: 'In certain countries they are saying that only birth control provides a solution to the problem. Only capitalists, the exploiters, can speak like that; for no one who is conscious of what man can achieve with the help of technology and science will wish to set a limit to the number of human beings who can live on the earth. . . . That is the deep conviction of all revolutionaries. What characterized Malthus in his time and the neo-Malthusians in our time is their pessimism, their lack of trust in the future destiny of man. That alone is the reason why revolutionaries can never be Malthusians. *We shall never be too numerous* however many of us there are, if only we all together place our efforts and our intelligence at the service of mankind, a mankind which will be freed from the exploitation of man by man.[12]

In such phrases not only does the well-known tendency of the Cuban revolution to voluntarism find expression together with a rhetoric of affirmation; but there is also the tendency to answer the irrational fears of the imperialist oppressor with equally irrational hopes. A materialist analysis of concrete needs, possibilities and limits, cannot be replaced by figures of speech. The Chinese leadership recognised that long ago and has therefore repeatedly modified its earlier population policy, which was very similar to the Cuban one in its premises. As far as the neo-Malthusians in the United States are concerned, a violent conflict has been raging for several years over their theses and their motivation.


## THE PROBLEM OF GLOBAL PROJECTION

A central ideological theme of the ecological debate as it is at present conducted — it is perhaps at its very heart — is the metaphor of 'space-ship earth'. This concept belongs above all to the repertory of the American ecological movement. Debates which are scientifically orientated tend to use a formulation which sounds more sober but, as far as content goes, comes to the same thing: they consider the planet as a closed and global ecosystem.

The degree of 'false consciousness' contained in these concepts is obvious. It links up with platitudes, which are considered to be 'idealistic' but to which even that word is misapplied: 'The good of the community takes precedence over the good of the individual', 'We are all in the same boat', and so on. The ideological purpose of such hasty global projections is clear. The aim is to deny once and for all that little difference between first class and steerage, between the bridge and the engine room. One of the oldest ways of giving legitimacy to class domination and exploitation is resurrected in the new garb of ecology. Forrester and Meadows, the authors of the MIT report, for instance,

by planning their lines of development from the start on a world scale, and always referring to the space-ship earth — and who would not be taken in by such global brotherliness? — avoid the need to analyse the distribution of costs and profits, to define their structural limitations and with them the wide variation between the chances of bringing human misery to an end. For while some can afford to plan for growth and still draw profits from the elimination and prevention of the damage they do, others certainly cannot. Thus, under accelerated state capitalism, the industrial countries of the northern territories of the world can maintain capital accumulation by diverting it to anti-pollution measures, to the recycling of basic raw materials, to processes involving intensive instead of extensive growth. This is denied to the developing countries which are compelled to exploit to the utmost their sources of raw materials and, because of their structural dependence, are urged to continue intensive exploitation of their own resources. (It is worth quoting in this connection the remark of a Brazilian Minister of Economics to the effect that his country could not have enough pollution of the environment if that was the cost of giving its population sufficient work and bread.)[13]

The contradictions which the ecological ideologies attempt to suppress in their global rhetoric emerge all the more sharply the more one takes their prognoses and demands at their face value. What would be the concrete effect, for instance, of a limitation of the consumption of energy over the whole of 'space-ship earth' such as is demanded in almost all ecological programmes?

Stabilisation of the use of energy – certainly, but at what level? If the average *per capita* consumption of a US citizen is to serve as a measure, then a future world society stabilised at this level would make an annual demand on the available reserves of energy of roughly $350 \times 10^{12}$ kilowatt hours. The world production of energy would then be almost seven times as great as at present and the thermal, atmospheric and radioactive pollution would increase to such a degree that the consequences would be unforeseeable; at the same time the available reserves of fossil fuel would disappear. If one chooses the present world average instead of the energy standard of the United States today as a measure of a future 'stable' control of energy, then the exploitation of the available source of energy and the thermal, chemical and radioactive effects in the environment would settle at a level only slightly higher than at present and one which would be tolerable in the long run. The real question would then be, however, how the available energy should be distributed globally. In arithmetical terms the solution would look something like this. The developing countries would have to have three times as much energy at their disposal as they do today; the socialist countries could by and large maintain their present level of consumption; but the highly industrialized countries of Europe and the USA would have to reduce their consumption enormously and enter upon a period of *contraction*.[14]

It must be clear that redistributions of such magnitude could be put through only by force: this is bound to hold good not only in international but also in national terms. Admittedly the captains of industry, gathered together in the Club of Rome, appear to have another view of conditions on board the ship in which we arc supposed to be sitting. They are clearly not plagued by doubts as to their own competence and qualities of leadership. On the contrary they assert that 'Very few people are thinking about the future from a global point of view.'[15] This minority leaves no doubt that they are determined to adjust their view of the world to suit their own interests. The scarcer the resources the more one has to take this view in distributing them; but the more one adopts this view of the world the fewer people can be considered for this high office.

Ecologists who find themselves confronted by objections of this kind will generally attempt to counter them by changing the terms of the argument. They will explain that their immediate task is to deal with a condition that exists in fact; this is a task that takes precedence over future distribution problems which it is not their task to solve. On a factual level, however, it is impossible not to treat the problem on a global scale; indeed it is inevitable. The pollution of the oceans or of the atmosphere, the spread of radioactive isotopes, the consequences of changes in climate caused by human intervention — all these are *actually*, and not merely in a ideological sense, world-wide and global phenomena and can be understood only as such.

While that is true, it does not help much. So long as ecology considered itself to be a branch of biology it was always conscious of the dialectical connection between the whole and the part; far from wishing 'merely' to investigate life on earth it saw itself as a science of interdependence and attempted to investigate the relations between individual species, the ecological sub-system in which they live and the larger systems. With the expansion of its research aims, its claims to hegemony and the consequent methodological syncretism, human ecology has forfeited that ability to differentiate which characterised its founders. Its tendency to hasty global projection is in the last analysis a surrender in the face of the size and complexity of the problem which it has thrown up. The reason for this failure is not difficult to determine. An ecologist researching the conditions of life in a lake has solid methodological ground to stand on; ecological arguments begin to become shaky only when ecologists involve their own species in them. Escape into global projection is then the simplest way out. For in the human case, the mediation between the whole and the part, between the sub-system and global system, cannot be explained by the tools of biology. This mediation is social, and its explication requires an elaborated social theory and at the very least some basic assumptions about the historical process. Neither the one nor the other is available to present-day ecologists. That is why their hypotheses, in spite of their factual core, are so easily overcome by ideology.

## ENVIRONMENTAL APOCALYPSE AS AN IDEOLOGICAL PAWN

The concept of a critique of ideology is not clearly defined — nor is the object it studies. It is not only that 'false consciousness' proliferates in extraordinary and exotic luxuriance given the present conditions under which opinions are manufactured, but it is also as consistent as a jellyfish and capable of protean feats of adaptability. So far we have examined the most widely diffused components of environmental ideology chiefly with regard to the interests which they at once conceal and promote. This would have to be distinguished from an evaluation in terms of an ideological critique which sees the ecological debate as a symptom that yields conclusions about the state of the society which produces it. So that nothing may be omitted, interpretations of this kind will now be briefly surveyed, although it is doubtful whether that will bring to light any new perspectives.

From this point of view, the preoccupation with ecological crisis appears as a phenomenon belonging entirely to the superstructure — namely an expression of the decadence of bourgeois society. The bourgeoisie can conceive of its own imminent collapse only as the end of the world. In so far as it sees any salvation at all, it sees it only in the past. Anything of that past that still exists must be preserved, must be conserved. In earlier phases of bourgeois society this longing for earlier cultural conditions was concentrated on 'values' which either did obtain previously or were believed to have done so. With the progressive liquidation of this 'inheritance', for example religion, the search for the roots of things, which is now thought to reside in what is left of 'nature', becomes radicalised. In its period of decadence the bourgeoisie therefore proclaims itself to be the protector of something which it itself destroyed. It flees from the world which, so long as it was a revolutionary class, it created in its own image, and wishes to conserve something that no longer exists. Like the apprentice sorcerer it would like to get rid of the industrialisation to which it owes its own power. But since the journey into the past is not possible, it is projected into the future: a return to barbarism, which is depicted as a pre-industrial idyll. The imminent catastrophe is conjured up with a mixture of trembling and pleasure and awaited with both terror and longing. Just as in German society between the Wars, Klages and Spengler sounded the apocalyptic note, so in the Anglo-Saxon lands today the ecological Cassandras find a role as preachers calling a class which no longer believes in its own future to repentance. Only the scale of the prophecies has changed. While Klages and Spengler contemplated the decline of Europe, today the whole planet must pay for our *hubris*. Whereas in those days a barbarian civilisation was to win terrible victories over a precious culture, today civilisation is both victim and executioner. What will remain, according to the prophecies, is not an inner but a physical desert. And so on. However illuminating such exegeses may occasionally sound, they cannot advance beyond a point of view that is little more than that of the history of ideas. Besides they do not carry much conviction in view of the fact that the dominant monopolies of the capitalist world show no signs of becoming aware of their presumed decadence. Just as German industry in the 1920s did not allow itself to be diverted from its expansion, so IBM and General Motors show little inclination to take the MIT Report seriously. Theories of decline are a poor substitute for materialist analyses. If one explores their historical roots it usually emerges — as in the case of Lukacs — that they are nourished by that very idealism which they claim to criticise.

## THE CRITIQUE OF IDEOLOGY AS AN IDEOLOGY

The attempt to summarise the Left's arguments has shown that the main intervention in the environmental controversy has been through the critique of ideology. This kind of approach is not completely pointless, and there is no position other than Marxism from which such a critical examination of the material would be possible. But an ideological critique is only useful when it remains conscious of its own limitations: it is in no position to handle the object of its researches by itself. As such it remains merely the interpretation of an interpretation of real conditions, and is therefore unable to reach the heart of the problem. Its characteristic gesture of 'unmasking' can turn into a smug ritual, if attention remains fixed on the mask instead of on what is revealed beneath it. The fact that we name the interests which lie behind current demographic theories will not conjure the needs of a rapidly growing population out of existence. An examination of the advertising campaigns of the enterprises involved does not increase the energy reserves of the earth by a single ton. And the amount of foreign matter in the air is not in any way reduced if we draw attention to the earlier history of pollution in the working-class quarters of Victorian England. A critique of ideology which is tempted to go beyond its effective limits itself becomes an ideology.

The left in West Germany has so far been scarcely conscious of this danger, or at least has not thought about it adequately, although it is by no means new in historical terms. Even Marxist thinking is not immune to ideological deformations, and Marxist theory too can become a false consciousness if, instead of being used for the methodical investigation of reality through theory and practice, it is misused as a defence against that very reality. Marxism as a defensive mechanism, as a talisman against the demands of reality, as a collection of exorcisms — these are tendencies which we all have reason to take note of and to combat. The issue of ecology offers but one example. Those who wish to deprive Marxism of its critical, subversive power and turn it into an affirmative doctrine, generally dig in behind a series of stereotyped statements which, in their abstraction, are as irrefutable as they are devoid of results. One example is the claim which is proclaimed in the pages of every other picture magazine, irrespective of whether it is discussing syphilis, an earthquake or a plague of locusts, that 'Capitalism is to blame!'

It is naturally splendid that anti-capitalist sentiments are so widespread, that even glossy magazines cannot avoid them altogether. But it is quite another question how far an analysis deserves to be called Marxist, which *a priori* attributes every conceivable problem to capitalism, and what the political effect of this is. Its commonplace nature renders it harmless. Capitalism, so frequently denounced, becomes a kind of social ether, omnipresent and intangible, a quasi-natural cause of ruin and destruction, the exorcising of which can have a positively neutralising effect. Since the concrete problem in hand — psychosis, lack of nursery schools, dying rivers, air crashes — can, without precise analysis of the exact causes, be referred to the total situation, the impression is given that any specific intervention here and now is pointless. In the same way, reference to the need for revolution becomes an empty formula, the ideological husk of passivity.

The same holds true for the thesis that ecological catastrophe is unavoidable within the capitalist system. The prerequisite for all solutions to the environmental crisis is then the introduction of socialism. No particular skill is involved in deducing this answer from the premises of Marxist theory. The question, however, is whether it adds up to more than an abstract statement which has nothing to do with political *praxis* and which allows whoever utters it to neglect the examination of the concrete situation.

The ideological packaging of such statements is dispelled at once, however, if one asks what exactly they mean. The mere question of what is meant by 'capitalism' brings to light the most crass contradictions. The comfortable structure of the commonplace falls apart. What is left is a heap of unresolved problems. If one understands by capitalism a system characterised by private ownership of the means of production, then it follows that the ecological problem, like all the other evils of which 'capitalism' is guilty, will be solved by nationalisation of the means of production. It follows that in the Soviet Union there can be no environmental problems. Anyone who asserts the contrary must be prepared to be insulted if a bundle of quotations from *Pravda* and *Izvestia* about the polluted air of the Don Basin or the filthy Volga is produced as evidence. Such a comparison of systems is forbidden — at least by Marxists like Gerhard Kade:

> For all those who are embarrassed by the question of the relationship between bourgeois capitalist methods of production and the destruction of the environment, a well-proven argument can be produced from that box of tricks where diversionary social and political tactics are kept. Scientists talk of comparing the two systems: standard common-place minds immediately think of the filthy Volga, the polluted air of the Don Basin or of that around Leuna. A whole tradition lies behind this. There is no social or political issue, from party conferences to reports on the state of the nation, where the diversionary effectiveness of such comparisons between systems has not already proved its worth. Whatever emerges from the increasing number of inquiries into environmental pollution in the socialist countries is dressed up scientifically and becomes a useful weapon in a situation where demands for replacement of the system begin to threaten those who have an interest in upholding present conditions. 'Go to East Germany if you don't like it here' or 'Throw Dutschke over the Wall' are the socially aggressive forms adopted by that diversionary maneouvre.[16]

Critique of ideology as ideology: the position which lays the blame on 'capitalism' is defended here at the cost of its credibility. Moreover the fact that in the socialist countries destruction of the environment has also reached perilous proportions is not even disputed, merely ignored. Anyone who is not prepared to go along with this type of scientific thinking is guilty of drawing analogies between the systems and is denounced as an anti-communist, a sort of ecological Springer. The danger that such a denatured form of Marxism will establish a hold on the masses is admittedly slight. The relationship of the German working class to its own reality is not so remote as to exclude the possibility of a comparative examination. In the face of such narrowness, one must

> bear in mind that capitalism as an historical form and as a system of production cannot be identified with the existence of a class of owners. It is an all-embracing social mode of production arising from a particular type of accumulation and reproduction which has produced a network of relationships between human beings more complicated than any in the history of man. This system of production cannot simply be done away with by dispossessing private capitalists, even when this expropriation makes it possible in practice to render that part of surplus value available for other purposes which it is not used for accumulation. The socialist revolution cannot be understood merely as a transfer of ownership leading to a more just distribution of wealth while other relationships remain alienated and reified. On the contrary, it must lead to totally revolutionized relationships between men and between men and things — that is to say, it must revolutionize the whole social production of their lives. It will either aim to transcend the proletariat's situation, of alienation, of the division between work and its profit, the end of commodity fetishism, or it will not be the socialist revolution.[17]

Only such a view of capitalism, that is as a mode of production and not as a mere property relationship, allows the ecological problem to be dealt with in Marxist terms. In this connection the categories of use value and exchange value are of decisive importance. The disturbance of the material interchange between humanity and nature is then revealed as the strict consequence of capitalist commodity production.[18] This is a conclusion which makes the ideological ban on thought unnecessary and explains why ecological problems survive in the socialist countries too. After all, the contradiction between use value and exchange value is not superseded any more than wage labour and commodity production:

> Socialist society has remained a transitional society in a very precise meaning of the word — a social form in which the capitalist mode of production, compounded with new elements, continues to exist and

exercises a decisive pressure on the political sphere, on relations between human beings and on the relationship between rulers and ruled.[19]

No less decisive is the pressure which the persistence of the capitalist mode of production exercises on the relationship between humanity and nature — a pressure which, on very similar lines to industrial production in the West, also leads to the destruction of the environment in the countries where the capitalist class has been expropriated.

The consequences of this position are extremely grave. It is true that it is possible in this manner to derive the catastrophic ecological situation from the capitalist mode of production; but the more fundamental the categories, the more universal the result. The argument is irrefutable in an abstract sense but it remains politically impotent. The statement that 'capitalism is to blame' is correct in principle, but threatens to dwindle into an abstract negation of the existing order of things. Marxism is not a theory that exists in order to produce eternal verities; it is no good Marxists being right 'in principle' when that means the end of the world.

Perhaps one has to remember that Marx represented *historical* materialism. From that it follows that the time factor cannot be eliminated from his theories. The delay in the coming of revolution in the overdeveloped capitalist lands is therefore not a matter of theoretical indifference. But that it was delayed does not in any way falsify the theory; for Marx certainly regarded the proletarian revolution as a necessary but not an automatic and inevitable consequence of capitalist development. He always maintained that there are alternatives in history and that the alternative facing the highly industrialised societies were long ago expressed in the formula: socialism or barbarism. In the face of the emerging ecological catastrophe this statement takes on a new meaning. The fight against the capitalist mode of production has become a race with time which humanity is in danger of losing. The tenacity with which that mode of production still asserts itself fifty years after the expropriation of the capitalist class in the Soviet Union indicates the kind of time dimensions we are discussing. It is an open question how far the destruction which it has wrought here on earth and continues to wreak is still reversible.

In this situation one must be relentless in critically examining certain elements in the Marxist tradition. First of all, one must examine to what extent one is dealing with original elements of Marxist thought or with later deformations of theory. Compared with the range of such questions the 'preservation of the classics' seems a trifling matter. Catastrophes cannot be combated by quotations.

To begin with one must examine critically the concept of material progress which plays a decisive part in the Marxist tradition. It appears in any case to be redundant in that it is linked to the technical optimism of the nineteenth century. The revolutions of the twentieth century have throughout led to victory in industrially underdeveloped countries and thereby falsified the idea that the socialist revolution was tied to a certain degree of 'ripeness' and to 'the development of the productive forces', or was actually the outcome of a kind of natural necessity. On the contrary it has been demonstrated that 'the development of the productive forces' is not a linear process to which political hopes can be attached.

Until a few years ago most Marxists accepted the traditional view that the development of the productive forces was by its nature positive. They were persuaded that capitalism, in the course of its development, would provide a material base which would be taken over by a socialist society – one on which socialism could be constructed. The view was widely diffused that socialism would be more easily developed the higher the development of the productive forces. Productive forces like technology, science, human capabilities and knowledge, and a surplus of reified labour would considerably facilitate the transition to socialism.

These ideas were somewhat mechanistically based on the Marxist thesis of the sharpening of the contradictions between the productive forces on the one hand and the relationships of production on the other. But one can no longer assume that the productive forces are largely independent of the relationship of production and spontaneously clash with them. On the contrary, the developments of the last two decades lead one to the conclusion that the productive forces were formed by the capitalist productive relationships and so deeply stamped by them that any attempt to alter the productive relationships must fail if the nature of the productive forces – and not merely the way they are used – is not changed.[20]

Beyond a certain point, therefore, these productive forces reveal another aspect which till now was always concealed, and reveal themselves to be destructive forces, not only in the particular sense of arms manufacture and in-built obsolescence, but in a far wider sense. The industrial process, in so far as it depends on these deformed productive forces, threatens its very existence and the existence of human society. This development is damaging not only to the present but the future as well and with it, at least as far as our 'Western' societies are concerned, to the Utopian side of communism. If nature has been damaged to a certain, admittedly not easily determinable, degree and that damage is irreversible, then the idea of a 'free society' begins to lose its meaning. It seems completely absurd to speak in a short-term perspective, as Marcuse has done, of a 'society of super abundance' or of the abolition of want. The 'wealth' of the over-developed consumer societies of the West, in so far as it is not a mere mirage for the bulk of the population, is the result of a wave of plunder and pillage unparalleled in history; its victims are, on the one hand, the peoples of the Third World and, on the other, the men and women of the future. It is therefore a kind of wealth which produces unimaginable want.

The social and political thinking of the ecologists is marred by blindness and naivete. If such a statement needs to be proven, the review of their thinking that follows will do so. Yet they have one advantage over the Utopian thinking of the left in the West, namely the realisation that any possible future belongs to the realm of necessity not that of freedom and that every political theory and practice — including that of socialists — is confronted not with the problem of abundance, but with that of survival.

WHAT ECOLOGY PROPOSES

Most scientists who handle environmental problems are not visible to the general public. They are highly specialised experts, exclusively concerned with their carefully defined research fields. Their influence is usually that of advisers. When doing basic research they tend to be paid from public funds; those who have a closer relationship with industry are predominantly experts whose results have immediate application. Most non-specialists, however, aim to achieve direct influence on the public. It is they who write alarmist articles which are published in magazines as *Scientific American* or *Science*. They appear on television, organise congresses, and write the best sellers that form the picture of ecological destruction which most of us have. Their ideas as to what should be done are reflected in the reforms promised by parties and governments. They are in this sense representative of something. What they say in public cannot decide how valid their utterances are as scientific statements; yet it is worth while analysing their proposals,

for they indicate where the lines of scientific extrapolation and dominant 'bourgeois' ideology intersect.

The Americans Paul and Anne Ehrlich are among the founders of
human ecology, and are still among its most influential spokesmen. In
their handbook on ecology they summarise their proposals under the
heading 'A Positive Programme', excerpts from which are extremely
revealing:

2. Political pressure must be applied immediately to induce the
United States government to assume its responsibility to halt the
growth of the American population. Once growth is halted, the
government should undertake to regulate the birthrate so that the
population is reduced to an optimum size and maintained there. It
is essential that a grass-roots political movement be generated to
convince our legislators and the executive branch of the government
that they must act rapidly. The programme should be based on what
politicians understand best — votes. Presidents, Congressmen, Sena-
tors, and other elected officials who do not deal effectively with the
crisis must be defeated at the polls and more intelligent and responsible candidates elected.

3. A massive campaign must be launched to restore a quality
environment in North America and to *de-develop the United States.*
De-development means bringing our economic system (especially
patterns of consumption) into line with the realities of ecology and
the world resource situation. . . . Marxists claim that capitalism is
intrinsically expansionist and wasteful, and that it automatically
produces a monied ruling class. Can our economists prove them
wrong?

5. It is unfortunate that at the time of the greatest crisis the
United States and the world has ever faced, many Americans,
especially the young, have given up hope that the government can be
modernized and changed in direction through the functioning of the
elective process. Their despair may have some foundation, but a
partial attempt to institute a 'new politics' very nearly succeeded in
1968. In addition many members of Congress and other government
leaders, both Democrats and Republicans, are very much aware of
the problems outlined in this book and are determined to do something about them. Others are joining their ranks as the dangers
before us daily become more apparent. These people need public
support in order to be effective. The world cannot, in its present
critical state, be saved by merely tearing down old institutions, even
if rational plans existed for constructing better ones from the ruins.

We simply do not have the time. Either we will succeed by bending old institutions or we will succumb to disaster. Considering the potential rewards and consequences we see no choice but to make an effort to modernize the system. It may be necessary to organize a new political party with an ecological outlook and national and international orientation to provide an alternative to the present parties with their local and parochial interests. The environmental issue may well provide the basis for this.

6. Perhaps the major necessary ingredient that has been missing from a solution to the problems of both the United States and the rest of the world is a goal, a vision of the kind of Spaceship Earth that ought to be and the kind of crew that should man her.[21]

This is not the only case of a serious scientist presenting the public with a programme of this kind. On the contrary. Page upon page could be used to document similar ideas. They can be seen as a consensus of what modern ecology has to offer in the way of suggestions for social action. A collection of similar statements would only repeat itself; and we will therefore confine ourselves to one further piece of evidence. The following quotation is from a book by the Swede, Gösta Ehrensvärd, a leading biochemist, in which he attempts a comprehensive diagnosis of the ecological situation. His therapeutic ideas are summarised as follows. 'We are not *compelled* to pursue population growth, the consumption of energy, and unlimited exploitation of resources, to the point where famine and world-wide suffering will be the results. We are not *compelled* to watch developments and do nothing and to pursue our activities shortsightedly without developing a long-term view.' The catastrophe can be avoided, he says,

if we take certain measures *now* on a global scale. These measures could stabilize the situation for the next few centuries and allow us to bring about, with as little friction as possible, the transition from today's hectically growing industrialized economy to the agricultural economy of the future. The following components of a crash programme are intended to gain time for the necessary global restructuring of society on this earth.

1. Immediate introduction of world-wide rationing of all fossil fuels, above all of fluid resources of energy. Limitation of energy production to the 1970 level. Drastic restrictions on all traffic, in so far as it is propelled by fluid fuels, and is not needed for farming, forestry and the long-distance transport of raw materials.

2. Immediate total rationing of electricity.

3. Immediate cessation of the production of purely luxury goods and other products not essential for survival, including every kind of armament.

4. Immediate food rationing in all industrial countries. Limitation of all food imports from the developing countries to a minimum. The main effort in terms of development policies throughout the world to be directed towards agriculture and forestry.

5. Immediate imposition of the duty to collect and re-cycle all discarded metal objects, and in particular to collect all scrap.

6. Top priority to be given to research on the development of energy from atomic fusion as well as to biological research in the field of genetics, applied ecology and wood chemistry.

7. Creation of an international Centre to supervise and carry through action around the six points listed above. This Centre to have the duty to keep the inhabitants of this earth constantly informed through the mass media of the level of energy and mineral reserves, the progress of research, and the demograph situation.[22]


## A CRITIQUE OF THE ECOLOGICAL CRASH PROGRAMMME

In their appeals to a world whose imminent decline they prophesy, the spokesmen of human ecology have developed a missionary style. They often employ the most dramatic strokes to paint a future so black that after reading their works one wonders how people can persist in giving birth to children, or in drawing up pension schemes. Yet at the conclusion of their sermons, in which the inevitability of the End — of industrialisation, of civilisation, of humanity, of life on this planet — is convincingly described if not proved, another way forward is presented. The ecologists end up by appealing to the rationality of their readers; if everyone would grasp what is at stake, then — apparently — everything would not be lost. These sudden about-turns smack of conversion rhetoric. The horror of the predicted catastrophe contrasts sharply with the mildness of the admonition with which we are allowed to escape. This contrast is so obvious and so central, that both sides of the argument undermine each other. At least one of them fails to convince. Either the final exhortation, which addresses us in mild terms, or the analysis which is intended to alarm us. It is impossible not to feel that those warnings and threats, which present us with the consequences of our actions, are intended precisely to soften us up for the conversion which the anxious preacher wishes to obtain from us in the end; conversely the confident final resolution should prevent us from taking too literally the dark picture they have painted, and from sinking into resignation. Every parish priest is aware of this noble form of verbal excess; and everyone listening can easily see through it. The result is (at best) a pleasurable *frisson*. Herein may lie the total inefficacy of widely distributed publications maintaining that the hour will soon come not only for humanity itself, but for all living forms. They are as ineffective as a Sunday sermon.

In its closest details, both the form and consent of the Ehrlichs's argument are marked by the consciousness (or rather the unconsciousness) of the WASP, the white Protestant middle-class North American. This is especially obvious in the authors' social and political ideas: they are just as *unwilling* to consider any radical interference with the political system of the United States as they are *willing* to contemplate the other immense changes which they spell out. The US system is introduced into their calculations as a constant factor: it is introduced not as it is, but as it appears to the white member of the middle class, that is to say in a form which has been transformed out of recognition by ideology. Class contradictions and class interests are completely denied: the parliamentary mechanism of the vote is unquestionably considered to be an effective method, by means of which all conceivable conflicts can be resolved. It is merely a question of finding the right candidate and conducting the right campaigns, of writing letters and launching a few modest citizens' activities. At the most extreme, a new parliament will have to be set up. Imperialism does not exist. World peace will be reached through disarmament. The political process is posed in highly personalised terms: politics is the business of the politicians who are expected to carry the 'responsibility'. Similarly, economics is the business of the economists, whose task is to 'draw up' a suitable economic system – this, at least, one has the right to ask of them. 'Marxism' appears only once, as a scarecrow to drive recalcitrant readers into the authors' arms. All that this crude picture of political idiocy lacks are lofty ideas: the authors are not averse to make good the lack. What is needed is a 'vision', since only relatively 'idealistic programmes' still offer the possibility of salvation. Since the need is so great, there will no lack of offers, and the academic advertising agency promptly comes up with the concept of 'Spaceship Earth', in which the armaments industry and public relations join hands. The depolitici-

sation of the ecological question is now complete. Its social components
and consequences have been entirely eliminated.

Concrete demands can now cheerfully be made. There is no danger
that they may be implemented with disagreeable consequences. A
brake on population increase, de-development of the economy, draconian rationing, can now be presented as measures which, since they are
offered in a spirit of enlightened, moral common sense, and are carried
out in a peaceful, liberal manner, harm no interests or privileges, and
demand no changes in the social and economic system. Ehrensvärd
presents the same demands in more trenchant, apparently radical terms
— those of the coolly calculating scientist. Like the Ehrlichs, his arguments are so unpolitical as to be grotesque. Yet his sense of reality is
strong enough for him to demand privileges for himself and his work —
that is to say, the highest priority for the undisturbed continuation of
his research. One particular social interest, if a very restricted one,
thereby finds expression: his own.

'Many of the suggestions', say the Ehrlichs, 'will seem "unrealistic".
and indeed this is how we view them.'[23] The fact that not even the
authors take their own 'crash programme' seriously at least makes it
clear that we are not dealing with madmen. They reason why they seek
refuge in absurdity is that their competence as scientists is limited
precisely to the theoretical radius of the old ecology, that is to say, to a
subordinate discipline of biology. They have extended their researches
to human society, but they have not increased their knowledge in any
way. It has escaped them that human existence remains incomprehen-
sible if one totally disregards its social determinants; that this lack is
damaging to all scientific utterances on our present and future; and
that the range of these utterances is reduced whenever these scientists
abandon the methodology of their particular discipline. It is restricted
to the narrow horizons of their own class. The latter, which they
erroneously regard as the silent majority is, in fact, a privileged and
very vocal minority.

CONCLUSIONS: HYPOTHESES CONCERNING A HYPOTHESIS

There is a great temptation to leave matters there and to interpret the
forecast of a great ecological crisis as a manoeuvre intended to divert
people from acute political controversy. There are even said to be parts
of the Left which consider it a luxury to trouble themselves with problems of the future. To do that would be a declaration of bankruptcy;

socialist thinking has from the beginning been orientated not towards the past but towards the future. Herein lay one of its real chances of success. For while the bourgeoisie is intent on the short-term interests of the accumulation of capital, there is no reason for the left to exclude long-term aims and perspectives. As far as the competence of the ecologists is concerned, it would be a mistake to conclude that, because of their boundless ignorance on social matters, their statements are absolutely unfounded. Their methodological ineptitude certainly decreases the validity of their over-all prognoses; but individual lines of argument, which they found predominantly on the causality of the natural sciences, are still useable. To demonstrate that they have not been thought through in the area of social casues and effects is not to refute them.

> The ideologies of the ruling class do not reproduce mere falsifications. Even in their instrumental form they still contain experiences which are real in so far as they are never optimistic. They promise the twilight of the gods, global catastrophe and a last judgment; but these announcements are not seen to be connected with the identification and short-term satisfactions which form part of their content.[24]

All this applies admirably to the central 'ecological hypothesis' according to which if the present process of industrialisation continues naturally it will in the foreseeable future have catastrophic results. The central core of this hypothesis can neither be proved nor refuted by political discussion. What it says is of such importance, however, that what one is faced with is a calculation like Pascal's wager. So long as the hypothesis is not unequivocally refuted, it will be heuristically necessary to base any thinking about the future on what it has to say. Only if one behaves 'as if' the ecological hypothesis was valid, can one test its social validity — a task which has scarcely been attempted up to now and of which ecology itself is clearly incapable. The following reflections are merely some first steps along this path. They are, in other words, hypotheses based on other hypotheses.

A general social definition of the ecological problem would have to start from the mode of production. Everywhere where the capitalistic mode of production obtains totally or predominantly — that is to say, where the products of human labour take the form of commodities — increasing social want is created alongside increasing social wealth. This want assumes different forms in the course of historical development. In the phase of primitive accumulation it expresses itself in direct impoverishment caused by extensive exploitation, extension of working hours, lowering of real wages. In the cyclical crises, the wealth that has been produced by labour is simply destroyed — grain is thrown into the sea and so on. With the growth of the productive powers the destructive energies of the system also increase. Further want is generated by world wars and armaments' production. In a later phase of capitalistic develop-ment this destructive potential acquires a new quality. It threatens all the natural bases of human life. This has the result that want appears to be a socially produced natural force. This return of general shortages forms the core of the 'ecological crisis'. It is not, however, a relapse into conditions and circumstances from the historical past, because the want does not in any sense abolish the prevailing wealth. Both are present at one and the same time; the contradiction between them becomes ever sharper and takes on increasingly insane forms.

So long as the capitalist mode of production obtains — that is to say not merely the capitalist property relationships — the trend can at best be reversed in detail but not in its totality. The crisis will naturally set in motion many processes of adaptation and learning. Technological attempts to level out its symptoms in the sense of achieving a homeo-stasis have already gone beyond the experimental stage. The more critical the situation becomes the more desperate will be the attempts undertaken in this direction. They will include: abolition of the car, construction of means of mass transport, erection of plants for the filtration and desalination of sea-water, the opening up of new sources of energy, synthetic production of raw materials, the development of more intensive agricultural techniques, and so on. But each of these steps will cause new critical problems; these are stop-gap techniques, which do not touch the roots of the problem. The political consequen-ces are clear enough. The costs of living accommodation and space for recreation, of clean air and water, of energy and raw materials will increase explosively as will the cost of recycling scarce resources. The 'invisible' social costs of capitalist commodity production are rising immeasurably and are being passed on in prices and taxes to the depen-dent masses to such a degree that any equalisation through controlling wages is no longer possible. There is no question, needless to say, of a 'just' distribution of shortages within the framework of Western class society: the rationing of want is carried out through prices, if necessary through grey or black markets, by means of corruption and the sale of privileges. The subjective value of privileged class positions increases enormously. The physiological and psychic consequences of the environmental crisis, the lowered expectation of life, the direct threat from local catastrophes can lead to a situation where class can determine the life or death of an individual by deciding such factors as the availability of means of escape, second houses, or advanced medical treatment.

The speed with which these possibilities will enter the consciousness of the masses cannot be predicted. It will depend on the point in time at which the creeping nature of the ecological crisis becomes apparent in spectacular individual cases. Even dramatic phenomena such as have principally appeared in Japan — the radioactive poisoning of fishermen, illnesses caused by mercury and cadmium — have not yet led to a more powerful mobilisation of the masses because the consequences of the contamination have become apparent only months or years later. But once, at any point in the chain of events, many people are killed, the indifference with which the prognoses of the ecologists are met today will turn into panic reaction and even into ecological rebellions.

There will of course be organisational initiatives and political consequences at an even earlier stage. The ecological movement in the United States, with its tendency to flee from the towns and industry, is an indication of what will come, as are the citizens' campaigns which are spreading apace. The limitations which beset most of these groups are not fortuitous; their activity is usually aimed at removing a particular problem. There is no other alternative, for they can only crystallise around particular interests. A typical campaign will, for example, attempt to prevent the siting of an oil-refinery in a particular district. That does not lead, if the agitation is successful, to the project being cancelled or to a revision of the policy on energy; the refinery is merely built where the resistance of those affected is less strongly expressed. In no case does the campaign lead to a reduction of energy consumption. An appeal on these grounds would have no sense. It would fall back on the abstract, empty formulae which make up the 'crash programmes' of the ecologists.

The knot of the ecological crisis cannot be cut with a paper-knife. The crisis is inseparable from the conditions of existence systematically determined by the mode of production. That is why moral appeals to the people of the 'rich' lands to lower their standard of living are totally absurd. They are not only useless but cynical. To ask the individual wage-earner to differentiate between 'real' and 'artificial' needs is to mistake the real situation. Both are so closely connected that they constitute a relationship which is subjectively and objectively indivisible. Hunger for commodities, in all its blindness, is a product of the production of commodities, which could only be suppressed by force. We must reckon with the likelihood that bourgeois policy will systematically exploit the resulting mystification — increasingly so, as the ecological crisis takes on more threatening forms. To achieve this, it only needs demagogically to take up the proposals of the ecologists and give them political circulation. The appeal to the common good, which demands sacrifice and obedience, will be taken up by these movements together with a reactionary populism, determined to defend capitalism with anti-capitalist phrases.

In reality, capitalism's policy on the environment, raw materials, energy and population will put an end to the last liberal illusions. That policy cannot even be conceived without increasing repression and regimentation. Fascism has already demonstrated its capabilities as a saviour in extreme crisis situations and as the administrator of poverty. In an atmosphere of panic and uncontrollable emotions — that is to say, in the event of an ecological catastrophe which is directly perceptible on a mass scale — the ruling class will not hesitate to have recourse to such solutions. The ability of the masses to see the connection between the mode of production and the crisis in such a situation and to react offensively cannot be assumed. It depends on the degree of politicisation and organisation achieved by then. But it would be facile to count on such a development. It is more probable that what has been called 'internal imperialism' will increase. What Negt and Kluge have observed in another connection is also relevant to the contradiction between social wealth and social poverty, which is apparent in the ecological crisis: 'Colonialization of the consciousness of civil war are the extreme forms in which these contradictions find public expression. What precedes this collision, or is a consequence of it, is the division of individuals or of social groups into qualities which are organized against each other.'[25]

In this situation, external imperialism will also regress to historically earlier forms — but with an enormously increased destructive potential. If the 'peaceful' methods of modern exploitation fail, and the formula for coexistence under pressure of scarcity snaps, then presumably there will be new predations, competitive wars, wars over raw materials. The strategic importance of the Third World, above all of those lands which export oil and non-ferrous metals, will increase and with it their consciousness that the metropolitan lands depend on them. The 'siege' of the metropolis by the village — a concept which appeared prematurely in the 1950s — will acquire quite new topicality. It has already been unmistakably heralded by the policy of a number of oil-producing countries. Imperialism will do everything to incite the population of the industrialised countries against such apparent external enemies, whose policy will be presented as a direct threat to their standard of living, and to their very survival, in order to win their assent to military operations.

Talk in global terms about 'Spaceship Earth' tells us almost nothing about real perspectives and the chances of survival. There are certainly ecological factors whose effect is global; among these are macro-climatic changes, pollution by radioactive elements and poisons in the atmosphere and oceans. As the example of China shows, it is not these over-all factors which are decisive, but the social variables. The destruction of humanity cannot be considered a purely natural process. But it will not be averted by the preachings of scientists, who only reveal their own helplessness and blindness the moment they overstep the narrow limits of their own special areas of competence.

> The *human* essence of nature first exists only for *social* man; for only here does nature exist as the *foundation* of his own *human* existence. Only here has what is to him his *natural* existence become his *human* existence, and nature becomes man for him. Thus *society* is the unity of being of man with nature — the true resurrection of nature — the naturalism of man and the humanism of nature both brought to fulfilment.[26]

If ecology's hypotheses are valid, then capitalist societies have probably thrown away the chance of realising Marx's project for the reconciliation of humanity and nature. The productive forces which bourgeois society has unleashed have been caught up with and overtaken by the destructive powers released at the same time. The highly industrialised countries of the West will not be alone in paying the price for the revolution that never happened. The fight against want is an inheritance they leave to all humanity even in those areas where humanity survives the catastrophe. Socialism, which was once a promise of liberation, has become a question of survival. If the ecological equilibrium is broken, then the rule of freedom will be further off than ever.

# Notes and References

CHAPTER ONE

1. K. Marx, *Economic and Philosophical Manuscripts* of 1844 (Moscow: Progress Publishers, 1974) p. 145.

2. K. Marx and F. Engels, *The German Ideology* (New York: International Publishers, 1947) p. 7.

3. Ibid. pp. 35 – 7.

4. Ibid. quoted from the original, German edition, and not in the English version, by A. Schmidt, *The Concept of Nature in Marx* (London: New Left Books, 1971).

5. K. Marx, *Capital* (London: Allen & Unwin, 1938) vol. I, pp. 156 – 8.

6. Mao Tse-tung, *On Practice: Selected Readings* (Peking: Foreign Language Press, 1967). p. 59.

7. *Manuscripts*, p. 98.

8. K. Marx, *Grundrisse* (Harmondsworth: Penguin, 1973) p. 409.

9. Ibid. p. 706.

10. Ibid. p. 410.

11. *Manuscripts*, p. 97.

12. *Capital*, vol. I, p. 335.

13. *Manuscripts*, p. 65.

14. Ibid. p. 91.

15. *Capital*, vol. I, p. 352.

16. *German Ideology*, p. 39.

17. L. Althusser, *For Marx* (London: Allen Lane, 1969) p. 170.

18. *Capital*, vol. II, p. 367.

19. K. Marx, *Selected Works* (Moscow: Marx Engels Lenin Institute, no date) vol. I, p. 114.

20. G. Lukacs, *History and Class Consciousness* (London: New Left Books, 1971) p. 24.

21. F. Engels, *Dialectics of Nature* (London: Lawrence & Wishart, 1940) p. 164.

22. Ibid. p. 172.

23. Ibid. p. 35.

24. Ibid. p. 35.

25. Ibid. p. 26.
26. Ibid. p. 38.
27. Ibid. p. 26.
28. F. Engels, *Anti-Dühring* (London: Lawrence & Wishart, 1969) p. 33.
29. *Manuscripts*, p. 90.

CHAPTER TWO

1. B. Hessen, 'The Social and Economic Roots of Newton's *Principia*', in *Science at the Cross Roads*, ed. N. Bukharin *et al.* (London: Kniga, 1931; reprinted London: Cass, 1973); R. K. Merton, *Science, Technology and Society in 17th Century England* (Belgium: St Catherine Press, 1938; reprinted New York: Fertig, 1970); and J. Needham, *Science and Civilization in China* (Cambridge University Press, 1954).

2. J. Needham, 'Science and Society in East and West', in *The Science of Science*, ed. M. Goldsmith and A. Mackay (London: Souvenir Press, 1964).

3. J. D. Bernal, *The Social Functions of Science* (London: Routledge, 1939).

4. J. D. Bernal, *Marx and Science* (London: Lawrence & Wishart, 1952) p. 49.

5. H. Rose and S. Rose, *Science and Society* (London: Allen Lane, 1969).

6. D. K. Price, *Government and Science* (New York University Press, 1954).

7. R. Gilpin, *Atomic Scientists and Nuclear Weapons Policy* (Princeton University Press, 1962).

8. R. Barber, *The Politics of Research* (Washington, DC: Public Affairs Press, 1966).

9. D. Schooler, *Science, Scientists and Public Policy* (New York: Free Press, 1971).

10. R. E. Lapp, *The New Priesthood* (New York: Harper & Row, 1968).

11. E. B. Skolnikoff, *Science, Technology and American Foreign Policy* (MIT Press, 1967).

12. D. Greenberg, *The Politics of Pure Science* (Harmondsworth: Penguin, 1969).

13. J. J. Salomon, *Politics and Science* (London: Macmillan, 1973).

14. C. F. Carter and B. R. Williams, *Industry and Technical Progress* (Oxford University Press, 1957).

15. Science Policy Research Unit, *Annual Report* (Sussex University Press, 1974).

16. A. Weinberg, 'Criteria for Scientific Choice', *Minerva*, 1, 2 (1963).

17. J. M. Levy-Leblond and A. Jaubert, *Critique et Autocritique de la Science* (Paris: de Seuil, 1973).

18. More recent analyses, even though they avoid a Marxist framework, have somewhat rectified this tendency — for example, S. Blume, *Towards a Political Sociology of Science* (New York: Collier-Macmillan, 1974).

19. Merton, *Science, Technology and Society*.

20. Hessen, 'Social and Economic Roots of Newton's *Principia*'.

21. M. Polanyi, *The Logic of Liberty* (London: Routledge, 1945).

22. Haldane Report, *Machinery of Government*, Cmd. 9230 (London: HMSO, 1918).

23. Ibid. para. 67(a).

24. Rothschild Report, *A Framework for Government Research and Development*, Cmnd. 1272 (London: HMSO, 1972).

25. R. Williams, 'Some Political Aspects of the Rothschild Affair', *Science Studies*, 3 (1973) pp. 31 – 46.

26. E. Teller, 'Can a Progressive be a Conservationist?', *New Scientist*, 45 (1970) pp. 346 – 8.

27. M. Horkheimer, *Eclipse of Reason* (New York: Columbia University Press 1947).

28. M. Horkheimer and T. W. Adorno, *Dialectic of Enlightenment* (London: Allen Lane, 1973).

29. J. Habermas, *Towards a Rational Society* (London: Heinemann, 1971).

30. H. Marcuse, *One Dimensional Man: Studies in the Ideology of Advanced Industrial Society* (London: Routledge & Kegan Paul, 1964).

31. W. Leiss, *The Domination of Nature* (New York: Brazillier, 1972).

32. Marcuse, *One Dimensional Man*, p. xv.

33. D. Joravsky, *Soviet Marxism and Natural Science* (London: Routledge & Kegan Paul, 1961).

34. L. Graham, *Science and Philosophy in the Soviet Union* (New York: Knopf, 1972).

35. R. Suttmeier, 'Party Views of Science: the Record from the First Decade', *China Quarterly* (Oct – Dec 1970).

36. J. S. Horn, *Away With All Pests* (London: Paul Hamlyn, 1969).

37. J. Needham, J. Robinson and I. Raper, *Hand and Brain in China* (London: Anglo–Chinese Education Institute, 1971).

38. Personal interiews (H.R.), Wuhan University, Sacu Tour, 1973.

39. Science for the People, *China Science Walks on Two Legs,* (New York: Avon, 1974).

40. Mao Tse-tung, *Selected Works* (Peking: Foreign Language Press, 1967) p. 375.

41. Ibid. p. 379.

42. M. Millionschikov, in *The Scientific and Technological Revolution: Social Effects and Prospects* (Moscow: Progress Publishers, 1972) pp. 13–28.

43. P. Kapitsa, 'Basic Factors in the Organisation of Science and How They are Handled in the U.S.S.R.', *Daedalus*, 102 (2) (1973) pp. 167–76.

44. J. Ravetz, *Scientific Knowledge and its Social Problems* (Oxford University Press, 1971).

45. N. Ellis, 'The Scientific Worker', Ph.D. thesis (University of Leeds, 1969).

46. J. Ziman, *Public Knowledge* (Cambridge University Press, 1968).

47. E. Shils, 'Faith, Utility and the Legitimacy of Science', *Daedalus*, 103 (3) (1974) pp. 1–16.

48. J. Monod, *Chance and Necessity* (London: Cape, 1972).

49. J. Ellul, *The Technological Society* (London: Cape, 1965).

50. T. Roszak, *Where the Wasteland Ends: Politics and Transcendance in Western Society* (London: Faber, 1973).

## CHAPTER THREE

1. A. Schmidt, *The Concept of Nature in Marx* (London: New Left Books, 1973).

2. K. Marx, *Introduction to the Critique of Political Economy* (Calcutta, 1904) p. 300.

3. Ibid. p. 293.

4. S. Tagliagambe, *Attualità del Materialismo Dialettico* (Rome: Riuniti, 1974) p. 179.

5. F. Engels, *Dialectics of Nature* (London: Lawrence & Wishart, 1940) Preface.

6. Tagliagambe, *Attualità del Materialismo Dialettico*, p. 186.

7. Ibid. p. 188.

8. K. Marx, *Capital*, vol. I, (London: Lawrence & Wishart, 1974) p. 77.

9. Marx, *Capital*, unedited (Florence, La Nuova Italia, 1969) vol. IV, p. 91.

10. K. Marx, *Grundrisse* vol. II, (Florence: La Nuova Italia, 1970) pp. 393 – 9.

11. Marx, *Capital*, vol. I, p. 477.

12. *Capital*, Italian edn, vol. IV, p. 79.

13. W. T. Knox, *Science*, 181 (1974) p. 415.

14. V. Fuchs, *The Service Economy* (New York: Columbia University Press, 1970) p. 109.

15. Marx, *Grundrisse*, vol. II, p. 405.

16. B. Commoner, *The Closing Circle* (New York: Knopf) p. 267.

17. K. Marx, *Critique of Policial Economy*, p. 279 – 80.

18. H. B. G. Casimir, 'The Ominous Spiral', *Studium Generale*, 24, 1460 (1971).

19. D. J. de Solla Price, *Little Science, Big Science* (Columbia University Press, 1963).

20. Working Group On Appointment Policy Report, CERN (31 August 1972).

21. G. Jona-Lasinio, 'Changes in Scientific Practice in Technological Society', unpublished manuscript (1972) p. 14.

22. Marx, *Capital*, vol. I. pp. 352.

23. Marx, *Critique of Political Economy*, p. 302.

24. H. Brooks, *Daedalus*, 102 (1973) p. 125.

25. R. J. Yaes, 'Physics from another Perspective. A Cynical Overview', duplicated (Memorial University of Newfoundland, St John's Newfoundland, Canada).

26. G. Morandi, F. Napoli and C. Ratto, 'Un indagine sociologica sulla ricerca in fisica dello stato solido', duplicated.

27. J. R. Cole and S. Cole, *Science*, 183 (1974) p. 32.

28. Yaes, 'Physics from another Perspective'.

29. S. E. Luria, *Science*, 180 (1973) p. 164.

30. M. Perl, *Science*, 173 (1971) pp. 1211 – 15.

31. Jona-Lasinio, 'Changes in Scientific Practice'.

32. T. S. Kuhn, *The Structure of Scientific Revolutions* (Chicago, University Press, 1962).

33. Marx, *Capital*, vol. IV, p. 72.

CHAPTER FOUR

1. H. Rose and S. Rose, 'The Radicalisation of Science', *The Socialist Register* (1972) pp. 105 – 32.

CHAPTER FIVE

1. K. Marx, *Capital*, vol. I (London: Lawrence & Wishart, 1974) p. 174.
2. Quoted in *Realtime*, 6 (1973).
3. G. Friedmann; quoted in E. Mandel, *Marxist Economic Theory* (London: Merlin Press, 1971) p. 183.
4. R. Boguslaw, *The New Utopians: a Study of System Design and Social Change* (Englewood Cliffs, N.J.: Prentice-Hall, 1965).
5. W. Fairbairn; quoted by J. B. Jefferys, *The Story of the Engineers* (London: Lawrence & Wishart for the AEU, 1945) p. 9.
6. W. H. Whyte, *The Organisation Man* (New York: Simon & Schuster, 1956).
7. K. Marx, *Critique of the Gotha Programme*, ed. C. P. Dutt (London: Lawrence & Wishart, 1938).


CHAPTER SIX

1. S. Rose, *The Conscious Brain* (London: Weidenfeld & Nicolson, 1973).
2. H. Osmond and J. R. Smythies, 'Schizophrenia, a New Approach', *Journal of Mental Science*, 98 (1952) pp. 309 – 15.
3. L. L. Iversen and S. P. R. Rose, *Biochemistry and Mental Disorder* (London: The Biochemical Society, 1974).
4. W. Sargent, *The Unquiet Mind* (London: Heinemann, 1967).
5. O. Sacks, *Awakenings* (London: Duckworth, 1973).
6. F. F. de la Cruz, B. H. Fox and R. H. Roberts, 'Minimal Brain Dysfunction', *Annals of the New York Academy of Science*, 205, entire volume (1973).
7. P. H. Wender, *Minimal Brain Dysfunction in Children* (New York: Wiley, 1971) pp. 20, 90 – 1, 94, 131.
8. J. M. R. Delgado, *Physical Control of the Mind: Towards a Psychocivilized Society* (New York: Harper & Row, 1971).
9. P. R. Breggin, *U.S. Congressional Record H.R.*, vol. 118, no. 26 (Washington, 1972).
10. V. Mark and F. Ervin, *Violence and the Brain* (New York: Harper & Row, 1970).
11. E. M. Opton, Documents circulated at Winter Conference on Brain Research, Vail, Colo. (1973).
12. L. S. Penrose, in *The Social Impact of Modern Biology*, ed. W. Fuller (London: Routledge & Kegan Paul, 1971).

13. J. Beckwith and J. King, 'The XYY Syndrome: a Dangerous Myth', *New Scientist*, 64, 923 (14 November 1974) pp. 474 – 6.

14. Desmond Morris, *The Naked Ape* (London: Cape, 1973).

15. R. Ardrey, *The Territorial Imperative* (New York: Dell, 1971).

16. See, for example, L. Tiger and R. Fox, *The Imperial Animal* (London: Secker & Warburg, 1972).

17. I. Eibl-Eibesfeldt, *Theology, the Biology of Behaviour* (New York: Holt, Rinehart & Winston, 1970).

18. V. C. Wynne-Edwards, *Animal Dispersion in Relation to Social Behaviour* (London: Oliver & Boyd, 1962).

19. P. P. G. Bateson, *Are Hierarchies Necessary?* (London: Brain Research Association, 1974).

20. K. Lorenz, *Civilized Man's 8 Deadly Sins* (London: Methuen, 1974).

21. R. Hofstadter, *Social Darwinism in American Thought* (Boston: Beacon Press, 1955).

22. B. F. Skinner, *Beyond Freedom and Dignity* (London: Cape, 1972).

23. M. Horkheimer, *The Eclipse of Reason* (New York: Columbia University Press, 1947).

## CHAPTER SEVEN

1. F. Galton, *Hereditary Genius* (London: Macmillan, 1869).

2. T. H. Huxley, *Emancipation – Black and White* (1865); quoted in G. M. Frederickson, *The Black Image in the White Mind* (New York: Harper & Row, 1972).

3. K. Pearson, *Natural Life from the Standpoint of Science* (1900) p. 46; quoted in D. K. Pickens, *Eugenics and the Progressives* (Nashville: Vanderbilt Press, 1968).

4. E. L. Thorndike, *Educational Psychology* (1929) p. 308; quoted in Pickens, ibid.

5. F. Lenz, in *Human Heredity*, ed. E. Baur, E. Fischer and F. Lenz (London: Allen & Unwin,1931) pp. 623 – 701.

6. A. R. Jensen, *Harvard Educational Review*, 39, 1 – 123 (1969).

7. R. Herrnstein, *Atlantic*, 228, 53 (1971).

8. W. Shockley, *Review of Education Research*, 41 227 (1971).

9. W. Shockley, 'Dysgenics, Geneticity and Raceology', *Phi Delta Kappan* (January 1971) p. 305.

10. Jensen was challenged on this point at the Cambridge debate, July 1970, and did not deny it (see note 11).

11. Cambridge Society for Social Responsibility in Science; the debate is reprinted in *The Biological Bases of Behaviour*, ed. N. Chalmers, R. Crawley and S. Rose (New York: Harper & Row, 1971).

12. H. Eysenck, *Race, Intelligence and Education* (London: Temple Smith, 1971).

13. See the publications of the Campaign on Racism, IQ and the Class Society in Britain and the Campaign Against Racism in the United States.

14. *Guardian* (10 May 1973).

15. This phrase has been used both by the men themselves and others, for example A. Flew, *New Humanist* (6 July 1973); *Guardian*, ibid.; B. Barnes, in a Third Programme talk with J. Ravetz (1972).

16. See, for example, in *New Scientist* (28 June 1973) p. 832.

17. A more extended treatment of the arguments presented here can be found in D. Layzer, *Cognition*, 1, 265 (1972); R. C. Lewontin, *Bulletin of Atomic Scientists*, 26 (3) 2 (1970); L. Kamin, *Heredity, Intelligence, Politics and Psychology* (New York: Wiley, 1974); S. Rose, *The Conscious Brain* (London: Weidenfeld & Nicolson, 1973); and K. Richardson and D. Spears (eds), *Race, Culture and Intelligence* (Harmondsworth: Penguin, 1972).

18. CRIQCS, *Racism, IQ and the Class Society* (London, 1974).

19. See, for example, A. R. Jensen, *Educability and Group Differences* (London: Methuen, 1973).

20. B. Lewis, from personal communication, and in London: *Scientific Reports System Research Ltd.* (1965).

21. See P. Watson, in *Race, Culture and Intelligence*; and D. F. Johnson and W. F. Mihal, *Proceedings On-Line Conference* (1972) p. 49.

22. B. S. Bloom, *Stability and Change in Human Characteristics* (New York: Wiley, 1964).

23. 'When we turn to intelligence, it may seem paradoxical that selection should ever favour the less intelligent, and consequently it may be difficult to reconcile the theories presented above with the possibility of any given racial group having lower genetic potential than others. *Yet it is easy to consider such possibilities.* If, for instance, the brighter members of the West African tribes which suffered the depredations of the slavers had managed to use their higher intelligence to escape, so that it was mostly the duller ones who got caught, then the gene pool of the slaves brought to America would have been depleted of many high IQ genes. Alternatively, many slaves appear to have been sold by their tribal chiefs; these chiefs might have got rid of their less intelligent followers. And as far as natural selection after the shipment

to America is concerned, it is quite possible that the more intelligent negroes would have contributed an undue proportion of "uppity" slaves, as well as being much more likely to try and escape. The terrible fate of slaves falling into either of these categories is only too well known; white slavers wanted dull beasts of burden, ready to work themselves to death in the plantations, and under those conditions intelligence would have been counter-selective. *Thus there is every reason to expect* that the particular sub-sample of the negro race which is constituted of American negroes is not an unselected sample of negroes, but has been selected throughout history according to criteria which would put the highly intelligent at a disadvantage. The inevitable outcome of such selection would of course be the creation of a gene pool lacking some of the genes making for higher intelligence' (our emphases; Eysenck, *Race, Intelligence and Education*, pp. 46–7. But contrast this with: 'As it is, most of the experimental and statistical observational work so far has been done by hereditarians, who have been much less prone to reply on non-empirical modes of proof'; Eysenck, *Race, Intelligence and Education*, p. 130.

24. 'It is known that many other groups came to the USA due to pressures which made them very poor samples of the original populations; Italians, Spaniards and Portuguese, as well as Greeks, are examples where the less able, less intelligent were forced through circumstances to emigrate, and where their American progeny showed significantly lower IQs than would have been shown by a random sample of the original population. Other groups, like the Irish, probably showed the opposite tendency; it was the more intelligent members of these groups who emigrated to the USA, leaving their less intelligent brethren behind.'; Eysenck, *Race, Intelligence and Education*, p. 47.

25. W. Bodmer, in *Race, Culture and Intelligence*, p. 83.

26. S. A. Barnett and J. Burn, *Nature*, 213, 150 (1967).

27. A. Globus, M. R. Rosenzweig, B. L. Bennet and M. C. Diamond, *Journal of Comparative and Physiological Psychology*, 82, 175 (1973).

28. A. R. Jensen, 'Kinship Correlations Reported by Sir Cyril Burt', *Behaviour Genetics*, 4 (1974) pp. 1–28.

29. Kamin, *Heredity, Intelligence, Politics and Psychology*.

30. L. Hudson; quoted in *The Biological Bases of Behaviour*.

31. J. Hirsch, 'Jensenism: the Bankruptcy of "Science" without Scholarship', *Educational Theory* (January 1975).

32. B. McGonigle and S. McPhilemy, *Times Higher Education Supplement* (13 Sep 1974) p. 13..

33. Lewontin, in *Bulletin of Atomic Scientists* (1970).

34. In his recent book *Educability and Group Differences*, Jensen claims to have found a statistical manipulation by way of which he circumvents the between/within group problem. However, such a manipulation scarcely touches the real issues involved; see, for example, De Fries, in *Genetics, Environment and Behaviour*, ed. L. Ehrman, G. Omann and E. Caspari (New York: Academic Press, 1972).

35. Jensen, *Educability and Group Differences*.

36. R. C. Lewontin, 'The Apportionment of Human Diversity', in *Evolutionary Biology*, ed. T. Dobzhansky, M. K. Hecht and W. C. Steere (New York: Appleton-Century Croft, 1972) vol. 6, p. 396; and M. W. Feldman and R. Lewontin, 'The Heritability Hang-up', *Science*, 190 (1975) 1163–8.

37. See Kamin, *Heredity, Intelligence, Politics and Psychology*; Pickens, *Eugenics and the Progressives*; and C. Karier, 'Testing for Order and Control in the Corporate Liberal State', *Educational Theory* (Spring 1972).

38. Pearson, *Natural Life from the Standpoint of Science*.

39. H. W. Holland, *Atlantic Monthly*, 52, 447 (1883); quoted in *Eugenics and the Progressives*.

40. H. Goddard, *Human Efficiency and the Level of Intelligence*: quoted in Ludmerer, *Genetics and American Society*.

41. L. Terman, 'The Conservation of Talent', *School and Society*, no. 483 (29 March 1924) p. 363; quoted in C. Karier, *Ideology and Evaluation* (Wisconsin Conference on Education and Evaluation, 1973).

42. Pickens, *Eugenics and the Progressives*.

43. K. M. Ludmerer, *Genetics and American Society* (Baltimore: Johns Hopkins, 1972).

44. R. Young, in *The Social Impact of Modern Biology*, ed. W. Fuller (London: Routledge & Kegan Paul, 1971).

45. See, for example, H. Rose and S. Rose, *Science and Society* (London: Allan Lane, 1969).

46. See Frederickson, *The Black Image in the White Mind*.

47. A. Montagu, *UNESCO Statement on Race*, 3rd edn (Oxford University Press, 1972).

48. A. M. Shuey, *The Testing of Negro Intelligence* (New York: Social Science Press, 1966).

49. In Britain in 1973 there were 280,000 black pupils, 200,000 of them children whose parents or grandparents were from the West Indies. Of this number, 300 were in grammar schools, 2500 in ESN schools. 31 per cent of ESN places in the Inner London Education Authority area are filled by black children, only 17 per cent of normal school

places. Even head teachers appear to believe that between 30 and 70 per cent of these black pupils are wrongly placed; *Evening Standard* (16 July 1973).

50. See S. Rose, *The Conscious Brain*.

51. Scottish Council for Research in Education, *Social Implications of the 1947 Scottish Mental Survey* (London University Press, 1953).

52. M. Skodak and H. M. Skeels, *Journal of Genetic Psychology*, 75, 85 (1949); H. M. Skeels, R. Updegraff, B. L. Wellman and H. M. Williams, *University of Iowa Studies in Child Welfare*, 15, 10 (1938).

53. Eysenck, *Race, Intelligence and Education*, p. 133.

CHAPTER EIGHT

1. A. Y. Lewin and L. Duchan, 'Women in Academia', *Science*, vol. 173 (1971) pp. 892–5; and N. Grunchow, 'Discrimination: Women charge Universities/Colleges with Bias', *Science*, vol. 168 (1970) pp. 559–61.

2. A common view in classical sociology was that women's social inferiority stems from the male's superior strength; see L. F. Ward, *Pure Sociology: A Treatise on the Origins and Spontaneous Development of Society* (New York: Macmillan, 1914) p. 349.

3. It is not without interest that Comte pointed out that the meaning of the concept 'family' is servants or slaves. In a not dissimilar vein Engels likens the relationship between husband and wife under capitalism as that of bourgeois and worker.

4. J. Mitchell, 'Women: the Longest Revolution', *New Left Review*, no. 40 (Nov–Dec 1966) pp. 11–37; and *Woman's Estate* (Harmondsworth: Penguin, 1970).

5. V. Packard, *The Sexual Wilderness* (London: Longmans, 1968).

6. A. Skolnick and J. Skolnick, *Family in Transition* (Boston: Little, Brown & Co., 1971).

7. Our own observations of communes would tend to support Abrams and McCulloch's dismal conclusions. However, some communes set up explicitly to help free the women and provide for the children, within the constraints imposed by the larger society, have achieved a certain modest success. P. Abrams and A. McCulloch, *Men, Women and Communes*, British Sociological Association Annual Meeting (1974).

8. *Employed* Russian women spent 415 minutes per day on household chores as against 185 minutes for *employed* men. The figures for the United States are: women 315 minutes; men 182 minutes. For Britain (London): 295 minutes; men 162 minutes. While the level of

domestic technology affects the over-all amount of household work to be done, the sexual distribution of housework, regardless of economic systems, seems to be that men do rather under 40 per cent and women 60 per cent; I. Cullen, *New Society*, 28 (601) (1974) pp. 63 – 5.

9. J. Belden, *China Shakes the World* (1949, republished Harmondsworth: Penguin, 1973).

10. W. Hinton, *Fanshen* (1966, republished Harmondsworth: Penguin, 1972).

11. We are grateful for discussions with several women interested in China, particularly Elizabeth Croll, for clarification of Chinese policy.

12. As the women's movement grows, Marxist groups find it increasingly necessary to develop both lines and activities which relate to women's liberation. At the 1974 Communist University in London the best attended section discussed 'Women's Liberation'. Earlier in the same year the *New Left Review* published W. Secombe's *The Politics of Housework: New Left Review*, 83 (1974) pp. 3 – 24.

13. L. Michel and C. Southwick, 'In Defence of Feminism', London conference report, mimeo (1972).

14. Mitchell, 'Women: the Longest Revolution'.

15. M. Dalla Costa and S. James, *The Power of Women and the Subversion of the Community* (Bristol: Falling Wall Press, 1972).

16. C. Lonzi, *Sputiamo su Hegel* (Milan: Scritti di Rivolta Femminile, 1970); also available in English, mimeo (London: Women's Resource Centre, 1972).

17. N. Himes, *A Medical History of Contraception* (Baltimore: William & Wilkins, 1936).

18. National Welfare Rights Organisation, 'Forced Sterilisation: Threat to Poor', *Welfare Fighter*, vol. 4, no. 1 (1974).

19. See, for example, the arguments put forward emphasising the interest of big US capital in population control in Latin America by Bonnie Mass, *The Political Economy of Population Control* (Montreal: Editions Latin America, 1972), and the ensuing replies published in *Science for the People*, 6 (2) (1974).

20. Institute of Biology, *The Optimum Population for Britain* (London: Blackwells, 1972).

21. P. Halperin, J. Kenrick and B. Segal, 'Fertility, Economics and Ideology', Women's Liberation and Socialism Conference, mimeo (London, 22 – 3 September 1973) pp. 71 – 82.

22. Ross Report, *Report of the Population Panel*, Cmnd. 5258 (London: HMSO, March 1973).

23. B. Ehrenreich and D. English, *Witches, Midwives and Nurses* (New York: Glass Mountain Pamphlets, undated).

24. Not only does the advent of scientific birth control increase the possibilities for women's freedom, but the actual technology itself is expressive of social relations between men and women. An example of the way in which the existing technologies are not neutral is exemplified in the difference between the condom and the cap or pill. The former is a male-controlled technology and the latter two female-controlled. The accounts in *Coal is Our Life* by N. Dennis, L. F. Henriques and C. Slaughter (London: Eyre & Spottiswoode, 1956) of the condoms being thrown on the fire because they diminish male satisfaction shows that where the technology is male-controlled, and there are conflicting interests — satisfaction versus pregnancy — it is the woman's interests which will be sacrificed. Thus the cap and the pill represent for women a gain in securing control over their bodies. We argue that future technologies, including those presently under discussion, are likely to be more repressive than liberatory.

25. Mary McCarthy's *The Group* (Harmondsworth: Penguin, 1970) in its portrayal of a small group of 1930s college-educated women, provides, in the account of Dorothy's visit to the birth-control clinic, a sensitive account of the kind of gentle feminist solidarity which many of these pioneering clinics provided.

26. G. Hawthorne, *The Sociology of Fertility* (London: Macmillan, 1970).

27. H. Suyin, 'Population Growth and Birth Planning', *China Now*, no. 43 (July –August 1974) p. 8.

28. A. Etzioni, 'Sex Control, Science and Society', in *Family in Transition*.

29. J. Postgate, 'Bat's Chance in Hell', *New Scientist*, vol 58, no. 840 (5 April 1973) pp. 12 - 16.

30. S. Rowbotham, *Women's Consciousness, Man's World* (Harmondsworth: Penguin, 1973).

31. W. Shockley, 'Dysgenics – A Social Problem: Reality Evaded by Illusion of Infinite Plasticity of Human Intelligence', *Phi Delta Kappan*, I (March 1972) pp. 291 - 5.

32. R. G. Edwards, B. D. Banister, and P. C. Steptoe, 'Early Stages of Fertilization in Vitro of Human Oocytes Matured in Vitro', *Nature*, 221 (1969) 632 - 5; also see a more general piece by R. G. Edwards, 'Aspects of Human Reproduction' in W. Fuller, *The Social Impact of Modern Biology* (London: Routledge & Kegan Paul, 1971) pp. 108 - 21.

33. J. B. S. Haldane in *Man and His Future*, ed. G. E. W. Wolsten-holme (London: Churchill, 1963) p. 337.

34. H. J. Müller, *Out of the Night* (New York: Vanguard, 1935).

35. W. Fuller, *The Social Impact of Modern Biology*.

36. In present-day society we can see trends in the mechanisation of reproduction which conceivably may reflect the transitional phase towards full biological engineering. As Margaret Stacey pointed out in a discussion of this chapter, for example, in one major Regional Health Authority, rather than allowing childbirth to take its natural course, there is a deliberate policy of inducing all births in order that children are born at hours convenient for the doctors.

37. K. Marx, *Capital*, vol. I (London: Allen & Unwin, 1938) p. 367.

CHAPTER NINE

1. M. Bookchin *Ecology and Revolutionary Thought*, in *Post Scarcity Anarchism* (London: Wildwood House, 1971) p. 11. Bookchin argues that to ask an ecologist exactly when the ecological catastophe will occur is like asking a psychiatrist to predict exactly when psychological pressure will so affect a neurotic that communication with him will be impossible.

2. *An Inquiry into the Sanitary Conditions of the Labouring Population of Great Britain*, Report from the Poor Law Commissioners to the Home Department, London, 1842, p. 68; quoted in J. Ridgeway, *The Politics of Ecology* (New York: Dutton, 1971).

3. Examples of this are not lacking in the ecology movement. In France there is an organisation for environmental protection which has an extremely right-wing orientation. The president of these 'eco-fascists' is none other than General Massu, the man responsible for the French use of torture in the Algerian war.

4. 'Profitschmutz und Umweltschmutz', in *Rote Reihe* (Heidelberg, 1973) 1, p. 5.

5. *Capital*, vol. I (Moscow: Progress Publishing, 1961) p. 510n.

6. Ridgeway, *Politics of Ecology*, pp. 22 – 5, sees Chadwick as an archetypal utilitarian bureaucrat, whose function was to secure the interests of capital by achieving peace and order among the poor. Better sanitation would produce a healthier and longer-living workforce. Sanitary housing would raise workers' morale, and so on.

7. Ibid. pp. 15ff; Ridgeway shows that over 150 years ago the Benthamites had evolved a theory of protecting the environment to

promote production. As he also points out, the measures taken in the advanced capitalist United States in the late 1960s failed to reach the standards of water and air cleanliness proposed by the utilitarians.

8. *Der Spiegel* (8 January 1973) p. 38.

9. Ridgeway, *Politics of Ecology*, pp. 207 – 11, analyses the 'eco-industrial complex', that is the growing role played by business in promoting ecological campaigns, such as Earth Day, and the liaison between business, politicians, local government and 'citizen campaigns'.

10. For illustration of the 'eco-industrial complex' in West Germany see 'Profitschmutz', p. 14, and the pamphlet *Ohne uns kein Umweltschmutz*.

11. 'Primera Conferencia de Solidaridad de los Pueblos de America Latina', in *América Latina: Demografía, Población indigena y Salud*, vol. 2 (Havana, 1968) pp. 15ff.

12. Ibid.

13. C. Koch, 'Mystifikationen der "Wachstumskrise" '. Zum Bericht des Club of Rome', *Merkur*, 297 (January 1973) p. 82.

14. G. Nebbia, *La Morte Ecologica* (University of Bari Press, 1972) 'Preface', pp. xv ff.

15. *The Limits of Growth, Report of the Club of Rome on the State of Mankind* (London: Earth Island, 1972) p. 13.

16. G. Kade, 'Kapitalismus and "Umweltkatastrophe" '; duplicated manuscript (1973).

17. R. Rossanda, 'Die sozialistischen Lander: Ein Dilemma des westeuropäischen Linken', *Kursbuch*, 30 (1973) p. 26.

18. Cf. 'Marx und die Oekologie', *Kursbuch*, 33 (1973) pp. 175 – 87.

19. Rossanda, 'Die sozialistischen Lander', p. 30.

20. A. Gorz, 'Technique, Techniciens et Lutte de Classes', *Les Temps modernes* (August – September 1972) 301 – 2, p. 141.

21. A. Ehrlich and P. R. Ehrlich, *Population, Resources, Environment* (San Francisco: Freeman, 1970) pp. 322 – 4.

22. G. Ehrensvärd, *Fore-efter. En Diagnos* (Stockholm: Bomnier, 1971) pp. 105 – 7.

23. Ehrlich and Ehrlich, *Population, Resources, Environment*, p. 322.

24. *Offentlichkeit und Erfahrung. Zur Organisationsanalyse von bürgerlicher und proletarischer Öffentlichkeit* (Frankfurt, 1972) p. 243.

25. Ibid. pp. 283ff.

26. Karl Marx, *Economic and Philosophical Manuscripts of 1844*, ed. D. Struik (Moscow: Progress Publishers, 1974) p. 137.

# Index